D0940293

ALONE AGAINST GRAVITY

Alone Against Gravity

Einstein in Berlin: The Turbulent Birth of the Theory of Relativity, 1914–1918

Thomas de Padova

Translated from the German by Michal Schwartz

B&B
Bunim & Bannigan Ltd.

Charlottetown-Montreal, Canada
www.bunimandbannigan.com

Originally published in Germany as
Allein gegen die Schwerkraft. Einstein 1914-1918

Bunim & Bannigan Ltd.
www.bunimandbannigan.com
P.O. Box 636, Charlottetown, PE,
C1A 7L3, Canada

QC16.E5P2513 2018 530.092 C2018901471-7
ISBN: 978-1-933480-47-3

Jacket design: Birgit Schweitzer, München, with images from: © akg-images.
Book design: Matthew MacKay
Printed and bounded in Canada on environmentally friendly paper by Marquis.

We gratefully acknowledge the translation of this work was supported by a grant from the Goethe-Institut.

Table of Contents

Introduction

This book is about the emergence of the general theory of relativity in the midst of World War I. It is about a catastrophe that spared no one and about a scientist in search of weightlessness. It reveals a world that is falling apart and, at the same time, how it is held together in an unparalleled way by Albert Einstein's physics.

Our tale begins on July 13, 1913, as Einstein is facing a decision that would change the course of his life. At the Zurich train station he meets with Max Planck and Walther Nernst, who have traveled from Germany to offer their significantly younger colleague a prestigious, high-profile post—a dream job without any teaching obligations at the Royal Prussian Academy of Sciences. Einstein must decide. Will he go to Berlin? To him, Planck and Nernst appear "like people who are out to acquire a rare stamp."[1]

The same date, July 13, 1913, also marks another daring new venture: after a clear night, at four o'clock, dawn, the Swiss pilot Oskar Bider climbs into his wooden flying machine and sets out to cross the entire Alps mountain range, from Bern to Milan. The motorized machine rolls on its bicycle tires over a pasture—and there is Bider, airborne, waving one more time at the curious spectators. He sets his course for the 3,500-meter-high Jungfraujoch.

While the marveling crowd looked up to Bider and his fellow pilots, celebrated heroes of the early twentieth century who had fulfilled one of humanity's oldest dreams, Einstein was turning our perspective on its head.

He was wondering: what would a person experience when free-falling to the ground from a high altitude? This scenario, which any pilot would call hell, offered the physicist an opportunity to observe the long established laws of gravity from a new angle.

If a person locked inside a capsule that is free-falling toward earth took his keys out of his pocket and let them loose, the keys would not hit the capsule's floor but remain suspended in midair. The person, too, would feel no gravitational force.

The notion of being weightless fascinated Einstein. He had told all his colleagues that gravity might not apply to someone who was free-falling. This was the core principle of his concept for a new theory of gravity that he had been developing for the past six years.

Exactly one year later, on July 31, 1914, just four months after Einstein's move to Berlin, Germany mobilized its army. Suddenly, science centers were filled with the roaring voices of nationalism.

Einstein wrote to friends in Zurich that nationalism was spreading like a dangerous disease, taking hold of otherwise capable and rational people.[2] Max Planck, Walther Nernst, and Fritz Haber, the very scientists who worked hard to bring the young genius to Germany, all joined in the war frenzy and descended into ardent nationalism.

As university rector, Max Planck exhorted students in Berlin to fight against "the hatcheries of insidious deviousness" against Germany.[3] As soon as the war began, Walther Nernst, a chemist, began exploring irritant gas as means of driving the enemy out of their trenches. The search for more potent chemical weapons began in the very institute where the director, Fritz Haber, offered his colleague Einstein a study room where he could peacefully ponder gravity and a general theory of relativity. While Einstein was tutoring Haber's twelve-year-old son in math, the institute's director left for the Western Front to prepare the deployment of poisonous gas.

Why did Einstein remain in Berlin, where he was in a top position as scientist but felt "alone, like a drop of oil on water, isolated by [his] attitude and cast of mind"?[4] Using parallel events, this book gradually guides the reader into Einstein's disrupted circumstances and his mind's cosmos. It presents the man and explorer as a contemporary witness of his time and shows how Einstein transformed from being a "pure" scientist and an apolitical man into being a politically engaged person and a pacifist by conviction.

The years between 1914 and 1918 were years of amazement and horror. They tell of the spirit of scientific optimism in Berlin and the unleashing of violent nationalism. They show a single researcher who allowed himself to be seduced by his own scientific questions but who also managed to stand up against inhuman rage.

The following chapters invite the reader to accompany Einstein to Berlin. Travel in time to the mecca of physics, where his idyllic research institute transformed step by step into a mega lab for weapons of mass destruction, where plans were forged for a war that not only employed airplanes and tanks for the first time, but placed all resources and inventions of the human spirit, from earplugs to uniform measurements of time, at the service of the military. "Our highly acclaimed technological progress, our civilization in general, is like an ax in the hands of a pathological criminal," Einstein declared bitterly in the midst of the war.[5] Naturally, he had no idea that a century later, his own highly abstract research would not only allow safer traffic navigation, but also enable the military to use drones and pinpoint missiles to their targets.

The general theory of relativity is Einstein's most significant scientific accomplishment. His linking of space and time, matter and gravity, introduced questions that occupy physicists and philosophers to this day. It has its own value. How we may benefit from it depends on whether we would follow Einstein's pacifist legacy.

Part I: Prelude

"The prelude is ruled by a lighter tone,
the supporting characters are the heroes."[6]

— Peter Sloterdijk

1. From Zurich to Berlin?

Over the Alps

Over the Alps now. The Swiss aviator's plan makes headlines. On July 13, 1913, as Oskar Bider climbs into his wooden flying machine in order to cross the Alps and fly from Bern to Milan, fifty curious bystanders surround the single-seater. They woke up early in order to witness the takeoff announced for four o'clock.

Following a starry night a fine mist is lying over the mountains. Will the wall of the Alps allow the young man with the white sweater and sport jacket to cross over? Will Bider fly his monoplane high enough to let the glacier saddle of the Jungfraujoch disappear below?

Many of the bystanders are wondering how the flimsy flying machine is supposed to carry him across the 3,500-meter-high saddle. Hot air balloons float effortlessly in the air, including a recent new construction by Count Ferdinand von Zeppelin, who in 1908 greeted Switzerland from up in the skies. His majestic airship, filled with hydrogen, made it lighter than air. In contrast, Bider trusts a flying machine that is evidently heavier than air. Still, this propeller-driven plane should gather enough speed in order to allow the airflow rushing over the wings to somehow generate the required lift; this puzzles even contemporary physicists.

The aeroplane underwent a rapid development. Not even ten years have passed since two bicycle manufacturers in the USA made an "air hop" for the first time, with a motorized biplane. Orville Wright lasted a full twelve seconds in the air, his brother, Wilbur, just short of one minute. When the Frenchman Louis Blériot crossed the English Channel in 1909, flying over the open ocean away from Calais and landing in Dover, aviation was suddenly on everyone's lips. Earlier that same year, a hundred thousand people thronged to flying competitions in the French Reims, Berlin, and the international exhibition of aviation in Frankfurt, where the brothers Wright, Blériot, and other pioneers of flight were renowned as celebrities.[7]

"Today, flying over straits and wide planes is no longer an unusual event," announced a Bern reporter in the summer of 1913,[8] referring to the Frenchman Marcel Brindejonc's latest fight over Europe from Paris to Warsaw in just one day then farther to St. Petersburg and over Stockholm and Copenhagen, arriving back at Paris—a route spanning approximately 4,860 kilometers.[9] This flight showed "that the airspace over flatland poses no obstacle for the efficient airplane."[10]

Highlands are another matter. "Only two years ago, when a Swiss aviator wanted to fly from Berlin to Bern, the common question was: will he get over the Hauenstein mountain pass? That pleasant flight found its rapid end over Hauenstein."[11] But what is the Hauenstein in comparison with the Jungfraujoch?

Just two week earlier, when Bider made his first flight over the Alps, his Blériot XI plane took him near the Jungfraujoch several times, but not high enough. After a three-hour flight the disappointed pilot was forced to fly back to Bern.[12]

Nevertheless, Bider refused to wait for a stronger engine. Instead, he prepared for his new flight attempt by further reducing the weight of his plane, replacing his seat with a lighter one, and tanking less fuel. As he wrote to his friend in Milan, he planned to make a stopover in Domodossola. "This way I need less fuel and oil and can reduce the machine's weight by forty kilograms."[13]

Now he puts a leather helmet over his cap, puts on his safety glasses, and throws his scarf over his chin, to defy the icy temperature that is awaiting him up there.

The weather forecast is favorable. A mechanic tightens the screws one more time, then, to the sound of the roaring engine, two bicycle tires and a smaller stern wheel roll over the meadow. They carry a fuselage built from ash with wooden wings braced to a tubular framework, clunking as it gathers speed in the field near Bern. At 4:07, to the crowd's applause, the machine goes up in the air. Craning their necks, everybody looks up to watch him climb higher and higher.

"What is going on?" the as yet unknown writer Franz Kafka asked, as he saw Louis Blériot circling above him in a flying machine for the first time. "Twenty meters above the ground there is a man in a wooden cage, fighting off an invisible danger, of his own free will. We, on the other hand, stand down below penned in and inessential, and watch him."[14]

Many aviators struggled unsuccessfully. Between 1908 and 1913 in Germany alone more than four hundred paid with their lives for the brief high-altitude euphoria. Around twice as many flying machines were destroyed.[15] Is Oskar Bider, who two days ago celebrated his twenty-second birthday, the next goner in a flying box?

The crowd breathed a sigh of relief at the successful start. A few curious spectators quickly climbed a hill in order to follow as long as possible the monoplane that was circling above their heads, gradually gaining altitude. Bider flew his aeroplane toward mountains that grew bigger as he approached.

An hour later the crowd could still hear the quiet whirring of the plane's engine. Then the sound vanished.

News from the Eiger Glacier station reached the city: at 6:07 Bider steered his machine over the glacier saddle of the Jungfraujoch and conquered the great wall of the Alps. For another brief half hour his subsequent flight over the big Aletsch Glacier could be seen. Finally came the news from foggy Domodossola of Bider's short stopover and continuing flight to Milan.

Bider's crossing of the Alps would go down as one of the most significant achievements in the history of aviation. Thanks to him, we discovered that even the highest mountain could be conquered from the air.[16]

Finally, Just Research

On this same July 13, 1913, an average-sized man, broad shouldered with a black moustache, is standing in the Zurich train station. Albert Einstein's gaze glides over the passersby until he finally sees the two men for whom he has been waiting. The thirty-four-year-old finds it flattering that the two scientists have traveled from Berlin to make him a job offer. The previous day he had staved them off, wanting to examine once more where his inner compass would point him.

The visitors are Walther Nernst and Max Planck. As a theoretical physicist, Einstein feels a special bond to the latter. For a time, Planck was so immersed in Einstein's theory of relativity that all his other studies were relegated to the back burner. One of the first to recognize the theory's fundamental importance, he helped rewrite the laws of mechanics accordingly, supervised dissertations on Einstein's theory of relativity, and advanced it among leading colleagues to win their recognition. It was also due to him that Einstein was nominated for the Nobel Prize this year—for the third time.

Einstein could not accept the Berlin offer merely on grounds of gratitude. For that reason he had asked the previous day for twenty-four hours in which to reflect, and suggested that Planck, a passionate mountain hiker, and his companion, Walther Nernst, both having arrived in Zurich with their wives, use the time for an excursion into the mountains.

Now he pulls out a white handkerchief and waves it to his new colleagues.

With this agreed upon sign, Einstein is accepting the offer. He has decided to go to Berlin. Although he distrusts the Prussian military's authoritarian state he wants to leave Switzerland in order to become a member of the Prussian Academy of Sciences.

Planck and Nernst are greatly relieved. For several months they have been meticulously preparing Einstein's appointment. As a bright science organizer in the international arena, Nernst had undertaken the delicate task of establishing the financial and institutional framework. Einstein should not only become an Academy member and professor at the university, but obtain a leading position in the still-to-be-founded Kaiser Wilhelm Institute of Physics.

The forty-nine-year-old chemist Nernst, known for his research on thermodynamics and as the inventor of the "Nernst lamp," pulled strings in the business world. In a confidential letter, the banker and industrial magnate Leopold Koppel agreed to fund Einstein's salary of 12,000 marks for a period of twelve years "in order to be able to offer the appointee a sufficient overall salary."[17] This remarkable generosity softened Einstein's reservations regarding the "imperious and touchy but not dishonest" Nernst.[18] Soon, Einstein would speak of "short, fat Nernst" in a friendlier manner, as an outgoing man who had a witty remark on hand for every occasion.[19]

While Nernst's eyes sparkled behind his glasses with excited expectation, Planck was reserved. The slim, lanky fifty-five-year-old physicist came from a socially and culturally conservative family of civil servants. It was only among family and close friends that he would occasionally thaw, as during the house concerts in his villa in the Grunewald forest, at which he himself played the piano. The presence of the venerable theorist made even Einstein pay attention to proper dress code and a serious conversational tone.[20]

A month earlier, Planck had stood up as secretary before the assembled Academy members in order to speak strongly in favor of electing Einstein for "the most distinguished scientific institute in the state."[21] He said that in his eyes, Einstein's new concept of time surpassed "in boldness everything that has been accomplished in speculative natural science, yes in philosophical theories of epistemology."[22] His theory gives the entire system of physics a new consistent character.[23]

Planck appreciated Einstein not just as a theorist of relativity. Einstein had been the first to prove the significance of the quantum hypothesis for atomic energy and molecular motion. And Einstein could also be considered a master in dealing with and deepening classical theory.[24]

Einstein was not too pleased that people in Berlin had turned him into a jack-of-all-trades. At this point, he had only one goal in mind: to generalize the existing theory of relativity in order to incorporate gravity into its framework. Thus he found himself on shaky ground; above all, the mathematical configuration of the new theoretical construct was highly demanding. "One thing is certain, that never before in my life have I been so bedeviled by anything. I have become imbued with great respect for mathematics, the more subtle parts of which I considered until now, in my ignorance, pure luxury!"[25]

Nevertheless, it is Planck who does not expect much. During their meeting in Zurich he inquired about Einstein's progress, but he could not relate much to his fundamental ideas or to his earlier mathematical construction, not to mention that Einstein's general theory of relativity had hardly any links to physically verifiable questions. Instead of encouraging, Planck tried to dissuade him.

"As an old friend, I must advise you against it [the general theory of relativity], since you will not break through it, and if you do, no one will believe you."[26]

Nevertheless, as they bid goodbye in Zurich, Planck promised Einstein that his work would be generously promoted in Berlin. When Planck promptly made good on his promise and advocated for the funding of a solar eclipse expedition to confirm the theory of gravity, he commanded some respect from Einstein, but in the meantime, Einstein complained to his friends about the less than enthusiastic reaction of physicists to his work on gravity. His main ideas were hardly accessible to anyone, not even to Planck, a fact he ascribed to the extreme conformity of German scientists.[27]

Einstein shared with many Swiss republicans distrust for the Germans. Since leaving his ancestors' land as an adolescent, among other things to avoid military service, his inner distance from the society of the German Reich had grown even bigger. At fifteen he broke out of the traditional education mold; being compelled and obliged to fit into that rigid system threatened to smother his curiosity. "What this delicate little plant needs, aside for stimulation, is mainly freedom."[28]

Without it, his thirst for knowledge would inevitably perish.

When in the fall of 1894 Einstein's parents moved from Germany to Italy to try their luck once again in forming a new company, their son, whom they believed had stayed behind for high school, suddenly appeared at their door in Milan. He had obtained a medical certificate as well as a recommendation letter from a teacher so that he could enter another school later, and off he went. His reckless plan was successful: he managed to convince his parents that he would be able to make the desired leap to the Polytechnic in Zurich by other means. Young Einstein's experience of freedom would define his entire life.

He actually did study later in Zurich, where his defiance of dictated authority grew even stronger. Working as a clerk in a patent office, he saved enough money to become a citizen of liberal Switzerland, marry his Serb classmate Mileva Maric in January 1903, and celebrate his 1905 scientific breakthrough.*

He subsequently traveled extensively in Europe. After spending a year in Prague, he went back to Switzerland with his wife and two sons, Hans Albert and Eduard, in the summer of 1912.

Zurich had become Mileva's home. Feeling like a dove that has returned to her cote, she was all the more unhappy with Albert's renewed plan to go abroad. Go away again, already? Just for the sake of a career? As a professor at the ETH (the Swiss Federal Institute of Technology in Zurich), Albert had a well-paid position. He enjoyed all the freedom he could think of—certainly more than he had back in Bern, where he was at the mercy of an eight-hour day at the patent office. Why didn't he simply reject the Berlin proposal?

Since hardly anyone in Zurich shared his theoretical concepts, Einstein secretly hoped to meet scientists that would join him on his conceptual forays into higher dimensions. The academic activities in Berlin attracted many talented people. The metropolis was a mecca

* Over the course of four months that year, March through June, he published several ideas that revolutionized science. Among them he described how light consists of particles called photons—the foundation of quantum physics and the idea that eventually won him the Nobel Prize.

of research, and with the founding of more Kaiser Wilhelm Institutes, was well on the way to consolidating its leading role in scientific research. If nothing else, Einstein expected that a collaboration with the astronomers in Berlin would support his theory of gravity.

Besides, he wished to rid himself of all the tiresome duties involved in teaching and other jobs. The fact that in recent years he had changed his workplace several times was not least because of bureaucracy. "The paper pushing in the office is interminable—all this, so it appears, in order to provide the gang of clerks in the government offices a facade of legitimacy," he complained in Prague.[29] Basically, he could do without an institution altogether. A theorist should keep everything in his head. All he really needed was a few books for his research.[30]

As much as the opportunity to exchange ideas with his colleagues appealed to him, Einstein found lecturing annoying. His students on the other hand were undeniably appreciative of the remarkably unconventional manner of their professor. "As he approached the lectern with his somewhat shabby clothes, wearing pants that were too short and an iron watch chain, we were at first skeptical," recalled Hans Tanner, one of Einstein's Zurich students. Instead of arriving with an elaborately prepared lecture, Einstein brought with him seldom more than a note the size of a calling card. He built his talk around two or three keywords, which he himself perceived as a "trapeze act."[31] "But with his unusual way of lecturing, he conquered our recalcitrant hearts after his opening sentences."[32] After the weekly colloquium, Einstein even invited his young students to join him at the Café Terrasse, in order to discuss actual research questions.

The Berlin delegation was aware of his needs. Einstein was so much immersed in his research that he "would happily renounce the large course of lectures that he dutifully reads."[33] The German researchers promised that he would be able to pursue his scientific studies in Berlin without any teaching duties. He would receive a full-time academic position, the one privileged position of the department of physics and mathematics. An appointment for life. As university professor he would have the right but not the duty to give lectures. An irresistible proposal.[34]

Berlin Also Means Elsa

Einstein was exuberant. "A great honor is bestowed upon me," he wrote to his cousin Elsa Löwenthal after he had come to an agreement with Planck and Nernst. He would move permanently to Berlin next spring. "I am already looking forward very much to the wonderful time we will spend together!"[35]

Five days later Elsa received another letter, in which Albert wrote that seeing her regularly was the nicest thing awaiting him in Berlin.[36]

Then another five days later—in which time he had already received mail from her—the next letter: he still cannot believe his luck, they will finally be together. "And one of the main things that I wish is to see you often, to walk around with you and to chat with you."[37]

Three effusive letters to his cousin within two weeks bears significant testimony that it was not just for the sake of science that Einstein wanted to move to Berlin. He had met with Elsa just over a year ago, after not having seen her for many years. They had known each other since childhood, as they played together in their parents' family homes. Now she was in her mid-thirties, divorced, and had two daughters.

Einstein, whose marriage had long been in crisis, became so fond of Elsa during the few days they spent together in Berlin "that I can hardly tell you."[38] When he reminisced about their outing together to the Wannsee, he was overjoyed. What a shame they did not live in the same city! At the time, he assessed the chance of being appointed in Berlin as fairly small. Indulging in memories, he had written affectionate letters to Elsa for a while before he surrendered to the inevitable: the yoke of his marriage.

A few months after his farewell letter of May 21, 1912, Elsa took the initiative to rekindle their correspondence. She congratulated him on his birthday and asked for his photograph, and Einstein promptly invited her to visit him in Zurich. He said he would give a great deal to be able to spend a few days with her—without Mileva, his "cross." Better still, he himself would come to Berlin in order to see her again.[39]

With their offer of the position in Berlin, Planck and Nernst barged right into Einstein's newly established courtship of Elsa. All of a sudden two passions coalesced into a promising life plan.

Einstein's correspondence was now free of any of the considerations that had prompted him to avoid Elsa in the past. He had explained to Elsa in his farewell letter "that it would turn against them both and others, should they grow closer to each other."[40] But now he wanted to follow his heart. His wife Mileva had no other option but to pack again and accompany him to Berlin.

Einstein's attitude toward his colleagues likewise shows how unswervingly he held to his course. Neither Planck, with his well-meaning advice, nor the expectation the Prussian Academy had of his fellowship could dissuade him from his goal to extend the theory of relativity. Nothing could shake his belief that he was on the right path with his reflections on physics and his mathematical drafts.

In Germany, Einstein was expected to initiate a new theory of matter at the interface of physics and chemistry.[41] What does the inside of an atom look like? How is it described mathematically? Einstein did not ignore the questions that were presented to him, but for him, Berlin was a place of desire, where he wanted to follow the stream of his own thoughts. He was going to the German capital "as an Academy man without any obligations, virtually as a living mummy," he wrote to one of his colleagues a few days after Planck and Nernst had left Zurich. "I'm already looking forward to this difficult career."[42]

Einstein's passion and confidence determined the tone of his letter to the Prussian Academy of Sciences in which he finally officially accepted the position in Berlin: "When I think that each working day shows me the flaws of my mental capacity, it is only with a certain trepidation that I can accept the high distinction intended for me. But what encouraged me to accept your promotion was the belief that all that can be expected of a person is that he devote himself with all his might to one good cause; and I truly feel capable of that."[43]

2. Scientist Couples

Make Way for Women

Mileva Einstein and Marie Curie's paths intersected in 1913. They met in the spring when Mileva accompanied her husband on a lecture tour to Paris, the city of libertines and women who discarded their skirts, going out in the modern trousers even when not riding a bicycle. The fact that Madame Curie accompanied the guests from Switzerland on their sightseeing increased the charm of the French metropolis.

After a wonderful time in Paris with all its riches, Mileva and Albert traveled back to Zurich and wrote a warm thank-you letter to the Nobel Prize laureate, inviting her to join them for hiking in eastern Switzerland. So in early August, Marie Curie journeyed with her two daughters, aged nine and fifteen, and a governess to Graubünden. There she met with Mileva, the wife of her highly esteemed physicist colleague. Like her, Mileva had intended at a younger age to lead a self-determined life and pursue a career in science.

Both Mileva and Marie had experience with the male academic world. In their adolescence years, it was almost impossible for women to attend university in Serbia or Poland, the Austro-Hungarian Reich or in Germany. Apart from some exemptions for a few female students, the authorities remained obstinate. Women's associations and international women's unions struggled for reforms in the education system, access to all the occupations, and women's suffrage, making their voices heard through lecture events and petitions. But they were endorsed only occasionally by prominent persons. The German author and playwright Ludwig Fulda wondered, "how can a modern man, who has earned this title, deny women's right and ability for academic studies?" His demand: "Make way for everybody!"[44]

Image 1: *Berlin 1916. "Make Way" for Sororities*

In a survey of German scholars from 1897 titled "The Academic Woman," Fulda was rather alone in this opinion. Few university professors spoke for a careful opening of universities, or better still, establishing a university for women. The opposition within institutions was too strong. Max Planck, for instance, who fifteen years later would appoint the physicist Lise Meitner as university assistant, was convinced in 1897 that nature itself has dictated a woman's vocation as mother and housewife. "Amazons are naturally inept in matters of the mind as well."[45]

Born in Warsaw, Maria Sklodowska went on to receive the Nobel Prize before Planck. The daughter of two teachers, she completed high school with distinction. She then made an unusual arrangement with her older sister: in order to enable her to study medicine in France, Maria Sklodowska would work as a governess. She had various positions as governess for six years before she followed her sister to Paris in the fall of 1891. She stayed with her sister, enrolled under the name "Marie" at the Sorbonne, and, with her sister's support, devoted herself to studying physics, which she completed as valedictorian, despite language difficulties. After the exams she married the physicist Pierre Curie, who had courted her persistently.

At the same time, women all over Europe were breaking new ground with similar life plans. Studying abroad was the only way to obtain higher education for many of them, including Mileva Maric, who grew up in Vojvodina, a border region in south Hungary that she herself called a "little country of bandits." Mileva's mother came from a Montenegrian family; her father, a Serb, being an administrator, mastered the German language. Mileva herself was raised bilingually, attended a Serbian girls' school, and obtained special permission to change to a boys' school. Finally, she left her homeland and headed for Zurich, where in 1896, at the age of twenty, she enrolled as the only woman at the Federal Polytechnic's department of mathematics and physics—the very section where Albert Einstein studied.

Mileva and Albert

Mileva was a quiet, serious student, small, thin, and with a dark complexion, who, as a result of a minor hip dislocation, limped slightly, a fact that did not deter her from talking walks beside the lake in Zurich and hiking in the mountains. Whenever she played music on free afternoons with her best friend, Helene Kaufler, who lived in the same guesthouse, Einstein would happily join them. He would play the violin, Mileva the tamburica, an instrument similar to the mandolin, and Helena the piano.[46]

After three years of slowly getting closer to each other, Mileva and Albert became a couple. Albert wrote ardent love letters to his "babe" and "minx," his "everything": "Without you I would lack self-discipline, the motivation to work, the joy of living—in short, a life without you is no life."[47] Life was no life without physics either, so in the same breath Einstein would rave about the kinetic gas theory. His sweet "little witch" and "beloved witch" shared his enthusiasm for research.

Before long, the hot-blooded student informed his parents that he wanted to marry his "gypsy." Pauline Einstein, from whom Albert inherited not only his superb, unruly head of hair, but also his derisive manners, was from the very beginning against consorting with an academic woman, who was three-and-a-half years older than him to boot. "You are ruining your future and blocking your path through life," his mother scolded. "She is like you, a book," she fumed, even before she met Mileva. "But you should find a wife."[48]

Albert's father thought a man could indulge in taking a wife only when he had a stable income. Albert found such a view of the relationship between husband and wife abhorrent. Instead he envisaged shaping his future career together with Mileva. He was happy to find in her a peer and independent partner but nevertheless gave her a full account of the hostile attitude of his parents in order to also describe in detail how energetically he had defended his love to her back home.

Einstein's willpower and limitless confidence impressed Mileva; she most likely envied his self-confidence. Both wrote a diploma thesis on heat conduction. Shortly afterward Albert passed his diploma exams as the worst among four male candidates. Mileva, who had already had to repeat her intermediate exam, failed. Like many in this course of study, she had failed mathematics.

As he himself had once ditched school, Albert was not particularly concerned about this setback. He had little doubt that Mileva would pass the re-examination. "I will be so proud, when I will perhaps have a little woman doctor as my sweetheart & myself still an absolutely regular man!" They would happily beaver away and have piles of money.[49]

Reality looked different. After Einstein passed his final exams his relatives stopped supporting him financially. And however much he applied for a post and plagued the academic world from the North Sea to the southern tip of Italy, with his mediocre diploma grades, the physicist, who would be much courted later, was unable to find an assistantship anywhere.

In the summer of 1901 Mileva failed the exams for a second time, at a point when she knew she was expecting an illegitimate child. Even in Switzerland this was considered disgraceful for women. In the meantime, Einstein hired himself out as a teacher for various positions, avidly devoured professional journals, wrote his first article, and participated boldly in current debates. But the scientist-to-be could not support a family.

Presumably to avoid public attention and not wanting to risk their future plans, Mileva moved back with her family. She gave birth there to a girl, Lieserl; for inexplicable reasons, Albert never saw his daughter. He did not travel to Vojvodina for the birth or in the following months. Then Lieserl mysteriously disappeared from their lives entirely. Despite intensive inquiries later, it remains unknown whether the little girl was placed in the hands of relatives or given up for adoption or perhaps died in early childhood. Whatever happened, it cast a shadow over Mileva for the rest of her life. Even though she soon married and gave birth to two sons, she never regained the excitement for science and the cheerfulness of her student years in Zurich.[50]

Marie and Pierre

Marie Curie's two pregnancies were entirely different. While expecting each daughter's birth, she experienced an extremely creative phase. She equally accomplished her double role as scientist and mother. After their first meeting, Einstein acknowledged her "sparkling intelligence" and described her as "an honest person, who was almost in over her head with her obligations and responsibilities."[51]

The greatest reward for her untiring laboratory studies was receiving the Nobel Prize for Physics in 1903, the first woman to do so. She and her husband Pierre received the award together for their exploration of natural radioactivity. At the time, neither of them held a university position. The Sorbonne immediately established one for Pierre Curie; Marie became a laboratory head. At the age of thirty-seven she obtained a regular salary for the first time.[52]

When the Nobel Prize for Chemistry followed in 1911, Pierre Curie had been dead for five years; he was crushed under the wheels of a horse-drawn carriage while crossing the street in Paris. Marie Curie, left on her own, had to struggle against social opposition. Among other things, her membership application to the Prussian Academy of Sciences was rejected. She was perhaps France's most famous female researcher, but her application dropped a bombshell on the Parisian scene.[53] It rained rejections, not only because she was a woman and Polish, but also presumably of Jewish ancestry. What is more, this woman, who preferred to deliver her lectures wearing a plain black dress, had, after her husband's death, a five-year affair with a married scientist and father of four, whom Einstein knew personally, the physicist Paul Langevin.

The French press turned this into a scandalous story. Public spite was focused on the unfaithful husband as well as the "radium Circe" who has seduced him. Even the *New York Times* published the love letters that Marie and Paul had exchanged and that Langevin's wife passed on to a journalist, whom the unmasked husband finally challenged to a duel. It went off bloodless but was not to remain the sole gunfight in the context of this *liaison dangereuse*. From the very beginning, Mileva and Albert Einstein followed the affair that was hotly discussed in expert circles.

As soon as the Nobel Prize committee in Stockholm found out about the affair, Marie Curie was notified that she would not have been awarded the Nobel Prize for Chemistry for the discovery of radium and polonium had the affair been known before, and she was urged not to accept the prize or travel to Sweden for the award ceremony, advice she ignored. Accompanied by her older daughter, Iréne, she traveled to Stockholm, where she made a courageous appearance.

Upon her return she became ill with a renal disease, possibly due to high radiation exposure during her research, and suffered a psychic breakdown that led to her hospitalization. She turned her back on Paris and public life. It was primarily thanks to the intensive care of a nurse and campaigner for women's rights in Britain, where suffragettes had in the meantime sparked a complete civil war for women's right to vote, that Marie Curie recovered to the point that she could think of mountain hiking again.[54]

Mileva and Albert planned the route for their hiking trip so that their own "polar bear," nine-year-old son Hans, and his peer, Eve, Marie Curie's younger daughter, would be able to cope. It led through a high mountain valley where the Inn River rushed through wild gorges and narrow escarpments, past picturesque scenes and steep summits, across the Maloja Pass down toward Lake Como. "One of the most beautiful routes one can take," raved the physicist.[55]

Mileva had special memories of this region. She had traveled to Lake Como for the first time as a twenty-five-year-old student, and Albert was waiting for her there with open arms. From Lake Como they sailed to Colico, took walks in blooming gardens, and from one day to the next found that the gorgeous spring had turned into the deepest snow, up to six meters high. They took a narrow sleigh, just wide enough for the two lovers, up to the Spülgen Pass. They wrapped themselves with blankets and shawls and drove through the cold, embracing each other tightly to keep warm. The coachman referred to Mileva as *signora*. After a four-hour drive they continued on foot through the white landscape. She was so happy that she hardly noticed the strain of hiking in the snow.

She reminisced wistfully about that trip, one of the few occasions she had her sweetheart entirely to herself. Since then, Albert had alienated himself increasingly, year by year. Unlike the Curies, the creative collaboration they had shared during the time of their studies had died away after the first pregnancy. Since then, to their mutual disappointment, there had been no second order companionship between them "based on shared logical experience and a fellowship of truth-seeking."[56]

As Mileva wrote to her friend Helena, Albert did not have much time for her after six years of marriage. Her husband had just become a professor at the Zurich University. "But what is to be done, the one gets the pearls, the other the box... I am hungry for love and would be beside myself just to hear a yes, that I almost believe evil science is the culprit."[57]

What once connected them had in the meantime separated them. Mileva told Helena that Albert worked incessantly and lived only for his science. When he spoke about urgent problems in physics, Mileva did not understand him. She could no longer interpret the mathematical signs in his notebooks. She did not know what was on his mind.

In the fall of 1913, when Einstein was invited to a convention of the Society of German Natural Scientists and Doctors in Vienna, Helena Savic could see for herself how big the distance between husband and wife had grown. During the convention Helena was often mistaken for Albert's wife, for Mileva was not walking next to her husband but was following him.[58]

In the meantime, Einstein avoided being alone with the woman he once loved such that he would never love another again. At night he no longer shared a bedroom with her. He stayed longer in bed in the morning, played occasionally with the children, and in his free time played music with others. As he wrote to his new secret companion waiting for him in Berlin, this was the only way he could still bear "living together." Mileva had transformed into a mirthless, humorless creature, who took no pleasure from life and who smothered the joy of others through her mere presence.[59]

Discussing Gravity

Albert and Mileva did leave together for Graubünden in August 1913, but it is indicative that Mileva does not appear even once in Eve Curie's memories of the joint mountain hike. Once more Mileva disappeared behind her husband, who talked incessantly about his science.

Although vacation postcards reveal that Mileva was part of the group, Eve mentions only Albert Einstein and his older son, Hans Albert, in addition to her mother, her sister, and a governess. It may be that Mileva could only occasionally participate in the hiking, since she had to look after three-year-old Eduard, who was often sick. Marie Curie obviously had a nanny to take care of her daughters.

In the past months Madame Curie had become acquainted with the suffragettes in Britain. Mileva would have happily learned more about it. But her husband had the floor, speaking with an accent and often searching for the right word. "If only my beak was better polished in French."[60]

Marie Curie's visit left Einstein feeling embittered. Unlike earlier, when he emphasized her enthusiasm, in a letter to Elsa he called the just recently recovered university professor a "cold fish," who expressed her feelings only when she grumbled about things she did not like. "And her daughter, is even worse—like a grenadier." She, too, is very talented, he correctly assesses.[61] Iréne Curie would later take over her mother's laboratory and likewise receive the Nobel Prize for Chemistry for the discovery of artificial radioactivity. As a fifteen-year-old with green eyes and close-cropped hair, she kept her distance from Einstein and other strangers.[62]

Her younger sister, Eve, was more personable. As she relayed in her biography of her mother, she enjoyed hiking in the mountains through larch forests and narrow gorges along raging water, and back again through dales and across lush meadows with a view of the alpine peak. In the meantime, her mother and Albert Einstein were entirely immersed in scientific discussions.

As Eve Curie wrote, "the young people enjoyed the trip enormously and led the way." At times they fleetingly caught a couple of odd sounding sentences, as when the mustachioed professor said to Marie Curie: "You understand, that I have to know exactly what happens to the occupants of an elevator when it plunges into a void!" The younger generation would burst into laughter at this expression of such moving concern. "How could they know that the assumed plunging elevator poses the main problem of the theory of relativity?"[63]

An elevator that falls into a void—Einstein loved such thought experiments, startling his companions with impossible situations. In his company one might suddenly find oneself in a "hiking adventure," noted a later companion: "Steep abyss are gaping, and one must climb down there on a slope, risking his neck. But then, gradually, astounding views open up."[64]

In this era, elevators were a favorite stage for scenarios in physics. Inspired by the completion of the 342-meter Eifel Tower in Paris, in 1895 the rocketry pioneer Konstantin Tsiolkovsky suggested building an even higher construction: a freestanding tower that would reach into space.

Einstein thought of a different model: a windowless box in space entirely isolated from the world outside, with experimenting scientists inside. Could the occupants determine by any means whether the box is situated in a gravitational field of a celestial body, such as earth? One would think: yes! All they need to do is let go of an object, for instance a set of keys, and see if it falls down. But it isn't that simple.

A few years later in Berlin, Einstein moved to Haberlandstrasse 5, a typical upper-class building with a concierge and iron elevator that lifted him to the fourth floor. Let us imagine, the scientist wants to explain his thought experiment to his cousin Elsa, who did not study physics or anything comparable. He brings a special scale from the Physics Institute that reacts immediately to the smallest change in weight. They enter the elevator together. Einstein sets down the scale and asks Elsa to step on: 60 kilograms.

Then he pushes the button for the fourth floor, and the elevator briefly accelerates before advancing at a steady pace. In this short acceleration phase Elsa feels the scale is pressing stronger against her legs. Indeed, Albert's special scale is no longer showing 60 kilograms but 66.

Albert has fun with the small experiment. As they are arriving at the fourth floor, he pushes G for the ground floor. Now the elevator accelerates briefly downward. Elsa feels lighter, and when she looks at the scale, it fleetingly shows only 54 kilograms. Losing weight can be so simple!

Having arrived at the ground floor, Albert pushes four again. During their ride upward he takes from his bag a metal-cutting saw. "I am now going to cut the elevator rope!" The scale will show no weight at all during the subsequent free fall in the shaft, and Elsa will feel no gravitational field. "When I drop the saw, it will not fall on our feet, it will float." Elsa is prepared to believe everything he says, as long as he does not get down to business with his threats.

While at the patent office in Bern, the physicist had hit upon the idea that a free-falling person would not feel his own weight. Today we are familiar with this situation from television footage of astronauts floating in a space station. This idea left a deep impact on Einstein. Later he would speak of it as "the happiest thought" of his life.

The occupants of a space station feel as if gravitational field does not exist, as if their laboratory is far away from all other celestial bodies in space; if the space station was a windowless black box, this feeling would be reinforced. Astronauts perceive neither their weight nor that of their plates and cups. If they turn their mug upside down, no coffee spills. Everything floats.

Of course, the space station orbits around earth. It constantly free falls downward—like a descending elevator. Since the space station has a high horizontal speed and the globe always curves away beneath it, it never reaches earth. Instead of striking the ground at a certain point, it revolves incessantly around the globe.

At the beginning of the twentieth century the idea of some kind of box in space was a fantasy about the future. But talking with the Nobel laureate Marie Curie, Einstein expanded on the story of the elevator and weightlessness—possibly in a similar manner as months later at a naturalists' conference in Vienna: "Two physicists, A and B, wake up from a deep sleep and notice they are in a black box with opaque walls, equipped with all their instruments. They have no idea where the box is located or whether it is moving or not." They can determine, though, that all the objects they drop accelerate with the same rate toward a certain direction, for instance, downward. What conclusion can they reach?

Physicist A concludes that the box is immobile and is on a celestial body. The dropped objects then fall in the direction of this body's gravitational field. Physicist B has a different view. He does not think a celestial body is nearby. Instead, he thinks an external force has seized the box and is pushing it upward with regular acceleration. "Is there a criterion for both physicists that would enable us to discern who is right?" Einstein keeps on asking. "We have no such criterion."[65]

During the hike with Marie Curie, he explained to her what experiments indicate that the two versions are equivalent to each other, and why the seemingly simple reasoning of physics opens new perspectives. Before taking her hiking in the mountains, had he warned the two time Nobel laureate of his constant flow of words, which did not stop even while he tried to converse in French?

The lifelong autodidact needed listeners like her in order to be able to unfold his ideas. In earlier years Mileva fulfilled this role, but now he was seeking the company of others. He explained to Madame Curie his outlines of a general theory of relativity, and doing so helped rearrange his thoughts. In between, the two of them wrote a postcard to Paul Langevin. And before the hiking came to an end in Lake Como on August 10, Einstein had thought through his lecture for the upcoming naturalist convention in Vienna. All in all, he found the days together in the mountains very refreshing.

A Bit of Freedom

On that same August 10, as King Carol of Romania rose from his seat during a gala dinner, an international audience was hanging on his every word with its telegraphs, telephones, and typewriters. The monarch wanted to put an end to a terrible chapter, over which the

London Times alone had published more than 150 editorials.[66] The brutish war was finally over, the redistribution of the Balkans sealed.

The prime ministers of Greece, Serbia, and Montenegro, as well as a delegate from Bulgaria, were sitting next to the king and the Romanian queen in his palace in Bucharest. Carol looked around, filled with pride over the contract he had just signed. He regarded the treaty as a personal victory, the result of his skillful diplomacy. His words would appear the next day in the newspapers: "It is with exceeding joy that I look around me at the gathered delegates from the Balkan states, who have just signed a peace treaty in the capital of Romania." This peace had been eagerly desired all over Europe. He, Carol, entertained a great hope that a new era of mutual trust and welfare for the Balkan Peninsula had just began. "I am holding fast to the conviction that the peace treaty will be long lasting."[67]

The peace treaty was welcomed from London to Paris and to Berlin. The German Kaiser himself stood by the resolution, although his only reliable ally, the Habsburg Monarchy, did not agree to the setting of the borders. According to the treaty, the territory of the Serbian kingdom had almost doubled in comparison to the 1912 division. The rulers in Austria-Hungary were worried about the aggressive Serbian nationalism that extended to the provinces of Bosnia and Herzegovina, which were annexed by the Habsburgs in 1908, as well as to Vojvodina in southern Hungary. Worse still, Serbian nationalism was supported by Russia. The feeling in Vienna was that of being held in the claws of Serbia and Russia.

The Balkan countries once belonged to the Ottoman Empire, which in 1683 threatened to expand and reach Vienna. This multinational Reich gradually broke apart. It disintegrated from the outer borders inward and lost one province after the other in North Africa and Southern Europe. Greece and Serbia seceded, later Montenegro, Romania, and Bulgaria.

In 1912, the young states perceived a unique, historic opportunity to drive the Turks away from the entire Balkan Peninsula. The trigger was Italy's sudden attack on the last Ottoman territory in North

Africa. The Italian-Turkish War over what later became Libya was the first conflict in which airplanes and zeppelins were deployed both to observe enemy positions and to drop bombs from the air.

The Ottomans had little with which to counter the modern Italian air force, the warships and submarines, large-caliber canons and automatic weapons. Those aligned with Filippo Tommaso Marinetti, poet and founder of the Futurist movement, celebrated the beauty of the new weapons, their speed and destructive force.[68] The Ottoman reign began to falter in the European mainland as well, where the military defeats of the Ottomans in Libya started a cascade of assaults. "It was only after the Italian attack that the Balkan states felt strong enough to take up arms," concludes the historian Christopher Clark.[69]

But could a few small monarchies endanger the all-powerful Ottoman Empire?

Surprised by the attack, the European press marveled at the alliance between the up-to-then feuding Balkan states and their swift military success. After one violent attack the Turkish army was forced to flee and from that point was constantly driven back to the Bosporus.

The commentators' tone changed during the course of the war. Mileva Einstein reported to her friend Helena Savic about the new Balkan euphoria in the European press. Some newspapers called the joint war of the Serbs, Bulgarians, and Greeks a "genuine people's war," a war of "national duty and honor." "Their fight is supported by an unlimited faith in its liberating mission."[70] Kaiser Wilhelm II spoke of a "legitimate triumphant progress" of the Balkan states. Now, finally, the Islamists would be driven out of Europe again. And in no way would he wish to stop this important development in world history.[71]

The German Kaiser's attitude led to considerable discord between Berlin and Vienna. Kaiser Franz Joseph I and his military command staff were alarmed by how rapid and drastic the changes were to the political geography in the Balkan Peninsula. In the meantime, Serbia wanted to expand its realm even further, to the Adriatic Sea. Vienna insisted on the independence of an Albanian state and demanded that its neighboring country immediately withdraw its troops from the occupied Albanian regions.

Mileva and Albert Einstein could read day by day in the newspapers how tense the atmosphere was. Some indicated that a war between Austria-Hungary and Serbia was imminent. "If only Austria would keep quiet," Albert wrote on December 1912 to Helena Slavic, whom he and Mileva had visited in Belgrade a few years earlier and whom he now addressed as "Serbian heroine." A conflict with Austria would be bad for Serbia, even in the case of a victory. "I think, though, that this saber-rattling has little significance."[72]

Actually the Austrian invasion failed to come, although the crisis continued to intensify. Lack of support from the German side prevented the Habsburg Reich from attacking. Together with England, the German Reich tried to bring the conflict under control. The two great powers struggled with the Balkans in order to find a solution for the Albanian question, until a peace treaty ending the Balkan War was signed in London in May 1913.

Barely had the delegates left London, when Bulgaria waged a new war. Although the country had emerged from the First Balkan War as a great winner, it was dissatisfied over the division of the lands. As Bulgarian troops attacked Greek and Serbian troops, they suddenly saw themselves as being surrounded by enemies on all sides. The allies that had fought together were now fighting each other in order to redistribute the spoils. In Bulgaria alone almost 100,000 soldiers fell victim to the Second Balkan War.[73]

The newspapers were now filled with terrible news. Although in the spring and summer of 1913 only scant information from the war zone reached the outer world, almost all the European newspapers reported terrible violence. The *Daily Telegraph* referred to the Balkans in the first days of July 1913 as a "European slaughterhouse." At the same time, the liberal newspaper *Vossische Zeitung* spoke of an "extermination war," and the *Neue Zeit* (*New Times*) of an internecine war that "made all the war excesses of the last five years grow pale in comparison."[74] While the press revealed in great detail the atrocities of the Balkan War, at the same time it conveyed the impression that such butchery would be inconceivable in civilized Central Europe.

Finally, Bulgaria laid down its arms.

The contract in Bucharest ended what had been until then "the most horrible war in the modern age," according to the American Carnegie Foundation, which had sent experts to the Balkan Peninsula that summer in order to investigate the course of the war.[75] What had played out in the Balkans in recent months could not be described in words. According to the final report, all parties to the war participated in terrible massacres. Civilians were deliberately murdered and entire villages razed to the ground in "ethnic cleansing."

While refugees roamed the country, on August 10, 1913, King Carol of Rumania was celebrated by the crowd in a triumphal procession through the decorated streets of Bucharest. The monarch saw himself as the winner of the war. At an evening banquet he pledged a new "era of mutual trust."[76]

But the Balkans remained a trouble spot. In September 1913, while Mileva and Albert Einstein spent a week visiting Mileva's parents in Vojvodina, where her Serbian grandfather arranged for his grandson to be baptized in the Greek-Orthodox church, the saber-rattling from Vienna was renewed. Since Albania's independence was once again at stake, Austria-Hungary finally gave its Serbian neighbor an ultimatum: within the next eight days, by October 17, 1913, Belgrade was to leave once and for all the disputed Albanian region. Should Serbia fail to remove its troops, Austria-Hungary would intervene with the "appropriate method" in order to enforce the request.

This time the threat did not fail to make an impression. Since the Serbian claims ran counter to the peace treaty just recently signed in Bucharest, Serbia could hardly count on support from Russia, and backed down. The ultimatum was celebrated in Vienna as a foreign policy triumph. Vienna was convinced that "in the end, Serbia understands only the language of violence."[77]

3. Metropolis

"Cars sped out of long, narrow streets into the shallows of bright squares. Pedestrians formed dark, cloudlike lines. Where stronger strokes of speed crossed through the otherwise casual rush, they thickened, trickled hastily, and after oscillating a little, resumed their steady pace. Hundreds of different tones entwined, unifying into a wiry texture of sounds with barbs protruding here and there, sharp edges running along it and submerging again, with clear tones splintering off and fading away. This sound, the peculiarity of which does not lend itself to description, would allow anyone who has been absent for years to recognize with closed eyes that he stands in the Reich's capital. Like people, cities can be recognized by their pace."[78]

— Robert Musil

The Capital of Technology

Berlin, Spring 1914. Viaducts and bridge piers slow the main traffic arteries. Under streets and squares, gangs of workmen force tunnels through the ground. The expansion of the above- and underground rail system progresses from station to station. Fifty train stations have been completed in the rail network since the turn of the century, giving the city an entirely new character. Every two minutes during peak times, trains leave on the main route, where department stores such as KaDeWe have opened.

Yet the traffic planning does not keep up with the influx of people. With more than three and a half million residents, Berlin has grown to be almost as big as Paris. While the last horse-drawn bus has already disappeared from the boulevards of the French metropolis, more people are transported in the German Reich's capital by horse-drawn buses than by motorized ones because "the pneumatic tires are limited to the first-class paved streets."[79] In the city's center, where over two million people dwell in the narrowest spaces, automobiles creep between carriages and wagons.

New residents describe Berlin as a city of noisy, whip-cracking coachmen, blaring klaxon honks, and the clanging of streetcars. In a letter in the *Berliner Tageblatt* (Berlin daily newspaper), one indignant reader notes that the noise of the streetcars is unbearable as their wheels screech around curves.

The pharmacist Maximilian Negwer formulated a tranquilizer for the frazzled residents of the metropolis: wax mixed with Vaseline and cotton and rolled into small balls for insertion into the ear. Negwer had been manufacturing earplugs in his "factory for pharmaceutical and cosmetic specialties" in Schöneberg since 1907 and supplied them to his artist friends around the Potsdamer Platz, one of the trendiest meeting places in Berlin.

Hustle and bustle dominated not only the Potsdamer Platz, "Europe's busiest square,"[80] but also the industrial quarters of northern and southern Berlin. The gate of the AEG, the Allgemeine Elektricitäts-Gesellschaft (General Electricity Company) in the Brunnenstrasse, led to a huge complex of administrative buildings, assembly halls, and state-of-the-art machinery. According to the contemporary author Arthur Fürst, "The walls of this 'city of electricity' enclose a population of more than 14,000 civil servants and workers.... On average, every two minutes a machine is completed, day and night."[81] Industrial, mass-produced goods, labeled as "Made in Germany," included an electric hairdryer that the AEG trademarked as "Foen" and an electric kettle, conceived by designer and architect Peter Behrens.

Electrical and mechanical engineering companies such as the AEG, Siemens & Halske, and Borsig manufactured light bulbs and lighting equipment, generators, and transformers in Berlin. Small engines and turbines, cars, locomotives, and airplanes. The companies had a considerable role in the redesign and electrification of the metropolis as well as in the economic upswing of the German Reich, which had risen to be the second biggest global commercial power after the British Empire. As the Swiss-French architect Le Corbusier concluded, "if Paris is the focal point of art, then Germany is the big production site."[82] The number of AEG employees alone doubled within five years from thirty-three thousand to sixty-six thousand.[83]

The economic power of Germany frightened many of its European neighbors. Wilhelm II did a few things to fan the flames of those fears. Just two years after the sinking of the *Titanic*, the biggest passenger liners and battleships were not big enough for the Kaiser. He kept the Reich's navy busy with his own sketches for cruisers and torpedo boats.[84] His enthusiasm for technology paired with his colonial and machtpolitik ambitions triggered a military alliance between former enemies, Great Britain, France, and Russia. The German Reich was feeling surrounded and became increasingly nervous.

The exceptional economic development of Germany could not mask the reactionary social order and constitution of the Reich. Wilhelm II scolded the Reich's parliament, calling it a "talking shop" and the "Reich's ape house" when its members debated the military budget, universal suffrage, or women's equality. He considered the domed building at the edge of Tiergarten park in Berlin far too majestic for the "bastards" in it, referring mainly to the members of the SPD (Social Democratic Party), which in 1890 was still illegal but had become the strongest faction, having won 34.8 percent of the vote in the 1912 elections. The Social Democrats actually won three quarters of the votes in the Reich's capital. The monarch's arrogance, the court's pomp, and the strong military presence in the capital did not go down well with Berlin workers, who were struggling to improve their miserable living conditions.

Boundless Enthusiasm for Flight

Top officials were happy that the Kaiser had never been able to stand being in Berlin for long stretches of time. The more he traveled, the fewer demands he made of them. At the beginning of the Easter holidays he was once again far away—on March 29, 1914, the imperial yacht arrived in Corfu.

Winter made an appearance this particular Easter weekend. Furs ruled again, noted a journalist who, despite the cold weather, went into the Grunewald forest, where the steeplechase season was opening, an event not to be missed by horse enthusiasts and those who made a point of attending every premiere.[85]

Yet the traditional horse race did not attract as a big a crowd as the appearance of a pilot at the other end of the city. Several thousand people took the streetcar or drove to the Johannisthal Airfield in the southeast of Berlin in order to watch the French aviator Adolph Pégoud, whose spectacular aerobatics on his last visit, half a year earlier, had thrilled spectators.

Aircraft manufacturers, airline companies, and aviation schools were all based in Johannisthal. Each of them profited from the public's fascination with flight. The French trailblazer's civil and military aviation inspired Prince Heinrich of Prussia, who appealed to

Image 2: *Looping the Loop: the attraction of Pégoud's Airshow*

the Germans in April 1912 to contribute to the building of airplanes and training of pilots. The "national aviation contribution" amassed 7,647,950 marks and 48 pfennigs, within six months.[86]

Melli Beese, the first female German pilot to make a name for herself, in Johannisthal in 1911, also hit the streets with a collection box for aviation. She had had a difficult time pursuing her interest in a male domain. Initially, no flight school would accept a woman, for women were considered unfit for flying just as they were for the military. At most they could hope to be taken on a passenger flight, provided they could pay for it.

When Beese was finally allowed to undertake flight training, she was harassed in every conceivable way. On her test she flew with an almost empty fuel tank because someone had almost completely drained it. "In the midst of the first flight curve, the engine suddenly stalled and she had to break off her flight immediately."[87] Beese opened her own flight school in 1912 and took credit for the fact that not a single serious accident occurred there.

Ever since the "national aviation contribution," the airfield had been under strong military influence. The Nobel Peace Prize laureate Bertha von Suttner bemoaned the fact that governments spent large sums on heightened armament and were appealing to the people's willingness to make sacrifices. In her prophetic essay on aerial bombardment, "The Barbarization of the Sky," she wrote: "The neighbor has an airship, ergo I must build one as well. The other neighbor has ordered two airplanes, hence I, too, must have two or possibly three."[88] Conquering the sky, one of mankind's most glorious achievements, was being mindlessly utilized for destructive purposes.

Many companies and fight schools in Johannisthal survived on military contracts. But the danger inherent in military flight was demonstrated in 1914 with the Prince Heinrich Flight, a competition in which the pilots were required to fly a route of almost one thousand kilometers in one day. Four officers died in a crash, and several pilots suffered severe injuries.[89]

Family Reunion

Albert Einstein arrived in Berlin on the Sunday of this Easter weekend, 1914. Unlike in later years, neither domestic nor international members of the press were lying in wait for him in the train station; instead, they were craning their necks in Johannisthal, scanning the crowd for prominent figures in the Grunewald, or reporting on the Australian magpie, a distant relative of the crow, that had just moved into the birdhouse at the zoological garden. The physicist arrived with a violin case and his—for most of his contemporaries—incomprehensible, unfinished theory of gravity. Its completion and experimental confirmation would make him world famous in a few years. It was in Berlin that Einstein would become a media star and icon of science.

The thirty-five-year-old had traveled alone by train. His wife and children would follow in three weeks. Mileva's leave-taking from Zurich had been delayed because Eduard, the youngest, had fallen sick with whooping cough and a middle-ear infection, and physicians recommended that he spend time in a sanatorium in Tessin.

Although Einstein made inquiries about Tete's "little earache" immediately upon his arrival, this delay suited him. As long as Mileva was taking care of the children in Switzerland, he could spend his days with his lover, undisturbed. He and Elsa had not seen each other for half a year and had bridged the long waiting period with letters. He had been courting his cousin, who called him "Albertle" in the Swabian dialect, by writing to her, though these letters were not necessarily the sizzling love letters he had sent Mileva fifteen years earlier.

Elsa possessed "nothing of the frailty and exoticism of her predecessor," as Einstein's biographer Jürgen Neffe points out. She was down-to-earth and strong-minded, open and sociable, and happily surrounded herself with people she could cook for.[90] Einstein imagined himself having a comfortable life with her, a union of coziness, peaceful nature outings, and scientific work. "Our walks in the Grunewald forest would be the loveliest, and in bad weather our getting-together in your small room."[91]

The weather was in fact bad upon Einstein's arrival, when Elsa received him. Their relatives knew about their affair. Elsa had already had her first row with Aunt Pauline, but things had calmed down. Albert's mother may have been good-natured, but as a mother-in-law she was a true devil, as he had to admit to his cousin.

Mileva certainly could have told Elsa a thing or two about that. She had cut off every contact with Pauline Einstein. Finally, on Christmas, Mileva became so upset with her mother-in-law's lack of tact that she sent back the packages intended for the grandchildren. Einstein, however, did not blame his mother for the rift, but Mileva. He told Elsa that his wife was the sourest sourpuss that had ever existed. "I shudder at the thought of seeing the two of you together. She will writhe like a worm if she sees you even from a distance!"[92]

Harsh words from a man who was courting a new partner. How did Elsa react, reading such lines? Would his emotional coldness frighten her?

We do not know, for only Albert's letters to Elsa have survived. In them he describes how much Mileva feared the company of his Swabian relatives. While he was happy to return to the family circle, Mileva felt oppressively surrounded by Albert's kin in her new residence. Now, of all times, her "devilish" mother-in-law had moved to Berlin because Albert's Uncle Jakob, her brother, had lost his wife in February. Until further notice, Pauline Einstein would run her brother's household in Charlottenburg, an affluent part of Berlin.

Since her husband's death, Pauline Einstein had been dependent on the family's support. She lived for a while with her sister Fanny and her brother-in-law Rudolf in Hechingen, then moved with them to Berlin, but she soon returned to Württemberg. At the age of fifty-six, she was now in the capital again, in order to give her brother a hand and be with her next of kin.

Fanny and Rudolf, who had prospered as a textile manufacturer, had rented a big apartment in Berlin-Schöneberg. In the same house, just one floor above, lived Elsa with her daughters Ilse and Margot, to whom Albert had already sent his greetings, referring to them as "little stepdaughters." In the meantime, Mileva got wind of her husband's affair with his cousin. A storm was brewing over Berlin. Mileva saw the dark clouds gathering yet did not want to risk a breakup with Albert, so quietly tolerated the affair.

Mileva had been to Berlin in January, for Albert had left it to her to look for a suitable apartment for the family, possibly near his future workplace in Dahlem. There, in southwestern Berlin, arable land had been converted for development since the turn of the century; as in Grunewald, villas and country houses for rich people had emerged. This is where Max Planck, the author Gerhart Hauptmann, and the big industrialist Walther Rathenau resided. Dahlem was to be developed not as a mere suburb in the country, but as a "German Oxford," a campus with the Reich's research institutions, university institutes, and the Kaiser Wilhelm Society for the Advancement of Science, which is succeeded today by the Max Planck Society.

Science in the Period of Wilhelminism

Science enjoyed a high status in the Reich's capital. The Urania Scientific Theater's slide- and movie-lectures attracted around 200,000 spectators a year. The mainly middle-class public embarked on a staged journey "from the earth to the moon" and followed, astonished, how one hundred wireless light bulbs were supplied with power or how a big, meters-long Tesla coil could send lightning-like electrical charges across the experimental hall. This is where Einstein, too, would present his new physics.

Things were a bit calmer in secluded Dahlem. Two years earlier, Wilhelm II had ceremoniously inaugurated the first two Kaiser Wilhelm Institutes there, including the Kaiser Wilhelm Physical Chemistry and Electrochemistry Institute directed by Fritz Haber, where the director reserved a room for the new Berliner, Albert Einstein. Haber was impressed by the clarity and depth of Einstein's theories. He had already consulted the ten-years-younger physicist on scientific matters, supported his appointment, and offered him a workplace in his institute, which both considered an interim solution but which marked the beginning of a close relationship between the two researchers and their families.

Haber lived in Dahlem with his wife, Clara, and eleven-year-old son, Hermann, in a magnificent executive villa with a big garden and "a whole apartment for visitors on the upper floor."[93] Mileva was happy about this haven. While she was looking for an apartment in Berlin, she stayed with the Habers, who helped her find an apartment in a three-story house on Ehrenbergstrasse at the corner of Rudeloffweg.[94] The spacious modern apartment, just a ten-minute walk to Albert's future workplace, had electricity and a telephone connection. The secondary school, where after the Easter holiday their older son, Hans Albert, would attend the first year, was easily accessible by foot.

As Einstein arrived in Berlin, the new apartment was still being painted. Until the renovation was completed and the mover had transported the household goods and furniture from Zurich to Ber-

Image 3: *Einstein's first workplace: The Haber Institute in Dahlem.*

lin, Einstein stayed with Uncle Jakob in Charlottenburg and was looked after by his mother. After a few days he sent a letter to a friend in Holland; he had not yet seen any physicists in Berlin other than Haber. He was truly delighted, though, with his local relatives, especially his cousin of the same age. "That is the main reason that I'm able to withstand this otherwise hateful big city."[95]

The pace and size of the metropolis were alien to Einstein. Furthermore, he felt uneasy about the spirit of the authoritarian state and the social snobbery of some of the residents of the city, including, Fritz Haber, his closest colleague for the time being. Since taking over the Kaiser Wilhelm Institute, Haber, a Prussian patriot with a typical "bragging scar" on his face from dueling as a student in Heidelberg, had been proudly wearing the title of "Privy Councilor."

"The German *Herr Geheimrat* (Mr. Privy Councilor) was a little god in those days," recounts the chemist Otto Hahn. "One had to deal with him very carefully and put up with his criticism, without being permitted to contradict him." Compared to England, the distance between a professor and his research assistants in Berlin was much bigger.[96]

Einstein was unhappy with Haber's need for recognition. He had to come to terms with the fact that this otherwise brilliant man was enslaved by vanity. This lack of personal tact was typical of Berliners. "When these people are in the company of French and English, what a difference!" The Berliners were coarse and primitive. "Vanity without self-awareness. Civilization (well-brushed teeth, elegant tie, smart moustache, impeccable suit), but no personal culture (coarseness of speech, movement, voice, feeling)."[97]

The researcher with the soft eyes had a sharp tongue. Even when writing about a colleague like Haber, Einstein did not mince his words. Although he was grateful for Haber's willingness to help, Einstein found his older colleague's ambition for power and perfectly styled appearance off-putting.

Most of all Einstein loathed the strong military presence in Berlin. Military parades were a daily occurrence in the garrison town; marching bands and children singing military songs were omnipresent. The middle class aspired after higher military ranks, although the class arrogance of the Prussian officers ran counter to the civilian system. "To serve as Prussian reserve officer—there were around 120,000 of them in 1914—was a much desired status symbol among the middle class."[98]

Fencing fraternity students appeared especially vigorous and dashing. Their self-imposed rules of conduct, codes of honor, and hierarchies were oriented to the military model. Student societies glorified the military actions of the German-French war of 1870–71 in a way that was similar to the almost three million members of the powerful Warrior Associations in the German Empire.

Einstein despised the Prussian uniform fetishism, and anyone who took pleasure in marching to music in rank and file was the object of his scorn. He wrote that such a person had been given a brain in error and could make do with just a spinal cord.[99]

Unfortunately, he couldn't avoid dress uniforms in Berlin. He bumped into large numbers of them at the gates of the Academy, which had been relocated to the Royal Library on the Unter den Linden boulevard. The world's biggest library at the time, at 170 meters long and more than 100 meters wide, the complex had opened just a week before Einstein's arrival after ten years of construction.

The German capital, a parvenu among European metropolises, exceled in science as well. Numerous big industrialists were involved in establishing the Kaiser Wilhelm Society for the Advancement of Science, among them the "cotton king," James Simon, who financed the 1914 excavation in Amarna, Egypt, and to whom Berlin owed thanks for the Nefertiti and Akhenaten busts; the coal entrepreneur Eduard Arnhold, who had just donated the Villa Massimo in Rome to the Prussian state as a cultural institute; and the banker Leopold Koppel, who financed the German-American professorial exchange and Einstein's position at the Prussian Academy of Sciences. All three of these men were of Jewish origin. "Strangely, the Jews here are undertaking German cultural activities, and the Germans respond with antisemitism," noted the poet Theodor Fontane.[100]

The fifty-nine-year-old Koppel in particular felt strongly about research funding. He was the founder of a private bank with a head office on the Place de Paris and principal shareholder in the Gas Light Company, known for its invention of the OSRAM lamp, whose factory site by the Spree river boasted Berlin's first high-rise building, eleven stories tall. With a donation of one million marks, Koppel, who was practically the sole funder of the Kaiser Wilhelm Institute for Physical Chemistry and Electrochemistry, helped the chemist Haber, whom he had tried to persuade to work for the corporation, to become the Institute's director. Now Koppel held out the prospect of building the Kaiser Wilhelm Institute for Physics. The banker surprised the designated director, Albert Einstein, with a "wonderful grandfather clock" as a welcome present and invited him to his home for the first week of April. In discussion with Koppel and Haber, Einstein would get a sense of the interweaving of capital and science in the Reich's capital.

The chemist Walther Nernst had also left a welcome greeting for Einstein. Nernst was preparing for a tour of South America, where he would deliver lectures at several universities. He was spending the Easter holidays at his château in Rietz some sixty kilometers away, where he would be studying Spanish with his wife. He told Einstein he would be back in the city on April 15.[101]

Like Haber, Nernst, who was the proud owner of a car, cultivated good financial contacts in industry. Several years earlier he had sold the patent for the Nernst lamp to the AEG, which owed its rapid rise to light bulb technology. Later Nernst would approach Telefunken, a wireless communications company, in order to market the first electromechanical piano, manufactured in a series, the "Neo-Bechstein." But he was not thinking only of profit; Nernst wanted to strengthen research by securing his students promising jobs in industry. In the department of chemistry, science and the economy were working together so closely and systematically that around 90 percent of the chemists with a doctoral degree were working in industry. Nernst and Haber strove for the same outcome in physical chemistry.

At the same time, Nernst espoused pure research. Together with industrialist Ernst Solvay he founded an international scientific summit that had been held two times in Brussels and that mainly served to bring quantum physics out of its shadowy existence. Haber, a rival of Nernst's, with neither the same head for theory nor up-to-date knowledge of quantum theory, was not accepted into this elitist circle; Einstein was. At the first Solvay Conference in 1911, the leading researchers had established that Einstein was in a class of his own. Everybody sensed that—especially Planck, Nernst, and also Haber. "The gentlemen from Berlin are betting on me as on a prize laying hen," Einstein told a colleague after a farewell dinner in Zurich. "But I do not know if I can still lay eggs."[102]

Thought Experiments on Rest and Mobility

At first the "gentlemen from Berlin" were satisfied with admiring the eggs that Einstein had already laid. Once, as it is reasonable to assume, an Academy member informed the *Vossische Zeitung* of Einstein's arrival and his hitherto pioneering work, the editor contacted him on April 14 and asked him to explain the theory of relativity in few words, a request Einstein was happy to fulfill. "Although a deeper insight into the theory of relativity cannot be obtained without considerable effort, even a layman will be excited to find out about some of the methods and findings of this new branch of theoretical physics."[103] Hence, just a few weeks after his arrival, the newspaper's readers were introduced to the physicist as the interpreter of his own theory.

In his article, Einstein presented the theory of relativity as the product of a historical process, which he wished to put across by means of daily experiences. Suppose we were sitting in a railway car and saw the train on the neighboring rail move past us. We might question whether we were moving or whether we were stationary while the other train rolled forward. "If we refrain from considering the shakings of the car, we have for the time being no way to determine whether the two trains are in fact moving."[104]

Now Einstein allows a physicist to board the car, equipped with all the possible measuring instruments. This time, the windows of the railcar are air- and light-proof. As long as the train drives along a straight stretch at a constant speed, the scientist cannot determine the direction of the train's movement or its speed. He cannot even assess whether it is moving at all. None of the experiments he conducted in the closed compartment could give him any indication.

The principle of relativity says that two identical closed systems are in no way different as regards the measurable occurrences inside them, as long as they move in a straight line and at the same speed in respect to each other. In physics such systems are called "inertial frames of reference." A train that suddenly brakes, accelerates, or turns a curve, does not form a frame of reference. Under such conditions we perceive changes—for instance, we are pressed into our seat, or a pencil lying on the table in front of us starts to roll.

Scientists did not really doubt the general validity of the principle of relativity as long as they could describe natural phenomenon with the laws of mechanics. But could one apply relativity on phenomena of a non-mechanical nature? Possibly on electrodynamics?

Einstein's childhood and youth were passed at a time of an upsurge in the electrotechnical industry, with the electrification of cities and of transportation. He was familiar with the hum of transformers from an early age, for his father Hermann and his Uncle Jakob operated the Electrotechnical Factory Einstein & Cie in Munich, which took a promising path with its production of small engines called dynamos and the installation of the city's lighting equipment in Schwabing. As the son, Einstein was expected to join the company.

However, it was not so much the technical devices that fascinated him but the physics of electrical currents and magnetic fields. Above all, he was excited by the theory of electromagnetism, established by the Scottish scientist James Clerk Maxwell. Maxwell's equations revealed how electrical and magnetic fields were interdependent. For instance, with the bicycle dynamo, the driving wheel is connected to a small magnet. When the bicycle moves, the magnet starts to rotate. The changing magnetic field induces a reaction from electrically charged particles in a wire. Those charge carriers move in a specific direction, generating an electrical current.

Maxwell's theory was exceptionally beautiful, but Einstein found in it one flaw: it handled a moving magnetic field and a stationary conductor differently than its opposite case, in which the magnetic field is stationary while the electrical conductors are moving. This irritated Einstein, who thought there should be no difference, because the generated current is always dependent on the relative motion. Why should there be two different explanations for the origin of an electrical current when a magnet and a coil are approaching each other? Einstein considered it worthwhile to extend the theory of relativity to electromagnetics.

With his electromagnetics, Maxwell opened a window to a hitherto unknown natural phenomenon. His equations enabled him to predict the existence of electromagnetic waves, the experimental evidence of which were finally reached by the physicist Heinrich Hertz. Hertz discovered radio waves and found out that they expand in the same way and just as fast as light. On closer examination, radio waves and light turned out to be two different types of electromagnetic waves. In 1901, when the Morse code letter "S" was sent over the Atlantic Ocean for first time with the help of radio waves and radio contact was accomplished, Einstein too experienced the beginning of a new, wireless transmission of information.

Opinions on the expansion of such electromagnetic waves were divided. Could they be transmitted without a carrier? On the eve of the twentieth century many physicists believed in the existence of an invisible ether, a secret substance that filled the entire universe. Light was supposedly propagated by this ether as sound waves in the air. Or was light, in the final analysis, formed by little particles that passed though the "empty" universe?

In the jungle of open questions, Einstein slowly groped his way to his own ideas. By the age of sixteen he had encountered a far-reaching question: what would it be like to pursue a light ray at near-light speed, traveling at 300,000 kilometers per second? Would light continue to be emitted at the same speed from this vantage point?

According to classic understanding, the speed of light should be slower. If we followed a light ray like a police car follows a traffic offender, the light ray would have to ultimately come to rest. Policemen perceive the driver of an overtaken vehicle as stationary and can gesture to him to pull over to the side of the road.

But would a fast-as-light observer actually perceive anything such as a resting light wave? Maxwell's equations did not allow that. In his youth Einstein was already convinced that a moving observer should perceive light the same as everyone else. The laws of nature must be independent of the motion of an observer. For many years he sought a way to expand the theory of relativity to phenomena such as light and bring it together with the insights gained on electrodynamics and mechanics.

The Relativity of Simultaneity

"The theory of relativity is established here," Einstein wrote in the *Vossische Zeitung*.[105] With it he had solved the seemingly contradictory situation while he was still Technical Expert Class 3 in the patent office in Bern. In 1905, as an outsider from scientific circles, he published a revolutionary work with the inconspicuous title "On the Electrodynamics of Moving Bodies" that today we refer to as the special theory of relativity.[106]

In the introduction Einstein referred to the already mentioned "intolerable" asymmetry in Maxwell's equations. He was convinced that rest and motion were only relative concepts. All laws of nature—not just those of mechanics, but also those of electrodynamics—ought to have exactly the same form in any inertial frame of reference. The second foundation of his special theory of relativity is that light is always propagated in an empty space with a constant velocity that is independent of the state of motion of the emitting body.

As explained above, these two fundamental assumptions appear to be incompatible. After intensive mulling, the young employee of the patent office discovered that all the discrepancies disappear if one is clear about the meaning of the concepts of time and space in a physical theory that encompasses not just mechanics, but also optical and electrical phenomena. Einstein's relativity physics is centered on a profound analysis of the concept of time.

What is it that we call time? What does it mean, in particular, when we say that two events take place at the same time?

"Until the arrival of the theory of relativity, it was believed that the statement 'two events occur at the same time in different places' had a certain sense, without having to define the concept of simultaneity in particular," Einstein explained to the readers of the *Vossische Zeitung*. "A more precise investigation, which does not dispense with a definition of simultaneity, shows that the simultaneity of two events is not absolute but instead can only be defined relative to an observer of a given state of motion."[107] In other words: simultaneity lies in the eyes of the observer.

It is easy to accept that the statement "at the same place" depends on the point of view of the observer. If I sit on a train, take a sip from my coffee mug, and place the mug back on the table, to me it is situated in the same place again. On the other hand, to an observer standing on the platform who sees me passing by on the train, the coffee mug has become quite remote from its original place. Always when observers move against each other, "sameness of place" is something relative.[108]

Einstein used concrete images to explore ideas. Moving trains that transformed into physics laboratories constantly reappear in his explanations. To illustrate the relativity of simultaneity, he again took up the perspective of an observer standing on a railroad embankment as a train goes by. This time something unusual happens: in Einstein's thought experiment the train is struck by lightning on both ends at precisely the moment when the observer is equidistant from each end of the train. Since the light from either end needs the same time span to reach the observer, he sees the two lightning strikes simultaneously.

The situation presents itself differently to a conductor standing in the middle of the train. Because the train is moving, the conductor is moving toward the light coming from the front while moving farther from the light in the rear. Since the speed of the light from either end is the same, the conductor sees the light in front slightly before the rear light. From his perspective the lightning occurred successively. "It follows that two events can happen simultaneously from the perspective of one observer yet happen at different times from the perspective of an observer that is moving in relation to the first," Einstein concludes. "This signifies a fundamental change in our concept of time."[109]

In his famous 1905 article, Einstein emphasized the importance of the simultaneity of events for our understanding of time. "We have to consider that all our judgments in which time plays a role are always judgments of simultaneous events."[110] For when we want to arrange an event chronologically, we must refer to another event that will give us a point of reference.

Let us put aside Einstein's physics for a moment. Our everyday language swarms with examples of such temporal relations: "I will call you when supper is ready." The timing of the contact is tied to another event: calling will occur when the food is ready.

Another example: "We are meeting today at midday." The event of our coming together coincides with another event, the middle of the day. And when is midday? Midday could be, for instance, when we feel hungry again. Midday could be when the sun reaches its peak. In our western culture, with its use of time zones and exact chronometers for defining temporal standards, synchronized watches have become a binding reference for everyone. To use Einstein's language, "We're meeting today at midday" means our meeting and the pointing of the clock's little and large hands to twelve are simultaneous events.

Generally, we can understand time as the relation between different events, while we use an event or an entire sequence of events as our standard.[111] When we do not want to follow the clock, we are using standards such as "when supper is ready." If, however, we want to catch a train, we must take the legal standard time and the difference between 12:00 and 12:01 seriously. Physicists like Einstein look at time as a measured quantity for calculating as accurately as possible the course of processes. Hence Einstein's abridged version: time is that which one reads on the clock.

"It would appear that all the difficulties pertaining to the definitions of 'time' can be overcome if instead of 'time' I use the position of the little hand on my clock." But according to Einstein, such a definition is sufficient only if "we define time just for the place where the clock is located; the definition will not be sufficient as soon as it connects events that occur in different locations."[112]

If we want to know whether an event on Earth and an event on the moon have taken place simultaneously, we can send a time signal from one to the other. No signal can travel faster than the speed of light. As a speed limit, the speed of light has a special significance in Einstein's physics. With a distance of almost 400,000 kilometers between Earth and the moon, it will take a light signal more than a second to reach the other celestial body. It takes about eight minutes for light to reach the sun. This duration has to be taken in consideration when synchronizing the earth-clock, the moon-clock, and a clock close to the sun.

It becomes more complicated when different observers look at events from referential systems that are moving at high speed relative to each other. As we saw, simultaneity to one observer seems non-simultaneous from the perspective of an observer who is moving in relation to him. They also register different durations of procedures. Their clocks tick at a different pace.

Imagine we are in a spaceship at the front of a fleet of spaceships. The onboard clocks are synchronized to each other. At exactly 13.00 we are flying at 87 percent of the speed of light past a space station, where the astronaut on duty has just sat on the bicycle-ergometer in order to keep himself fit. He carries a structurally similar clock, which he places on the exercise bike and sets to 13.00 at the moment we pass by. How much time will have passed before the astronaut gets off the bicycle after his workout?

In order to measure this duration, we have to read the clocks again. But since they are moving away from each other, the second synchronization cannot occur at the same place as the first. To estimate the pace of a clock that moves in reference to us, we always need a set of at least two synchronized clocks.[113]

Supposing that exactly at the moment the astronaut completes his training program and the clock on his bicycle shows 13.05, the last spaceship of our fleet passes by then and there. The crew looks at their onboard clock that is synchronized with our clock at the front of the fleet and determines that the hands have already moved to 13.10. From our perspective at this high relative velocity, the astronaut's clock is ticking half as fast as ours. And not just his clock ticks more slowly from our frame of reference. The movement of his legs on the bicycle and his heartbeat also appear delayed. Even the cigarette he smokes after his training burns down at half speed.

The key point is this: from his perspective he sees exactly the opposite. To him, his clock ticks completely normally, while the clocks in the space fleet are half as fast. The different perspectives are equally right. What is decisive is the relative velocity between the two inertial systems that in our example amounts to 260,000 kilometers per second.

It sounds contradictory to say that from the perspective of observers moving in respect to one another, each other's clock ticks more slowly. But the astronaut on the space station uses a different frame of reference and with it a different time standard. When starting the training program he looks at his clock, and through the ship's window at the clock in our passing spaceship. While he is cycling, our clock moves away from him. For a second clock's synchronization he has to rely on a clock that moves in unison with his own. He uses another temporal system of reference.

According to Einstein, time designation is meaningful only when we have a system to which we can refer this designation. In his newspaper article he describes this new time recording as the most important and contested part of his theory. It is outside our everyday life experience, since at our normal daily speed, time dilation is immeasurably small. The relativistic effect is directly visible only in processes that take place almost at the speed of light.

Imagine a light ray that bounces back and forth between two mirrors. The mirrors are one meter apart from each other, and their surfaces are perfectly parallel. The time the ray needs to bounce back and forth remains constant, and the time span is tiny. Since a light ray is emitted at around 300,000 kilometers per second, the time that elapses between two reflections is three hundred millionths of a second. In Einstein's world of ideas, such an arrangement provides an excellent time-measuring instrument. As a light clock, it illustrates with simple means why time is dilating in moving systems.

Light Clock

Let's assume that Einstein has set up his light clock so that the mirrors are parallel to the floor and the light ray runs up and down. Now a second observer is dashing past him at nearly the speed of light. From Einstein's perspective he is moving perpendicularly to the light's traveling direction.

Matters are different from the perspective of the second observer: during the 300,000,000th of a second that the light ray needs to reach the opposite mirror, the second observer moves almost a meter farther, and as a result, from his perspective, the light ray moves diagonally to the mirrors' surfaces. For him the light travels in a zigzag. Therefore the path is longer and so is the timespan between the two reflections. For according to the theory of relativity, the speed of light remains constant. Inversely, Einstein too sees the similarly structured light clock ticking more slowly as it passes by him.

In his newspaper article Einstein cannot expand on the extent to which time dilation challenges our usual understanding of time. Instead, he uses the opportunity to provide an outlook on his general theory of relativity. As already explained, the special theory of relativity relies on the assumption that rest and constant motion are physically equal. In inertial systems, the laws of nature have the same form for all observers. "The question then is whether the principle of relativity is limited to steady motion. Are perhaps the laws of nature constituted so that they are identical also for two observers who move unevenly in respect to one another?"[114]

Einstein's instinct tells him that accelerated motion has no special role. We have already familiarized ourselves with his first thought experiment, which harbors the germ of his general theory of gravity: an elevator that dives into a void. Meanwhile, the question of the relationship between space, time, and gravity remains open. But the passionate brooder is confident that he will soon be able to extract such a law of nature.

In order to be able to focus he has chosen what appears to be a

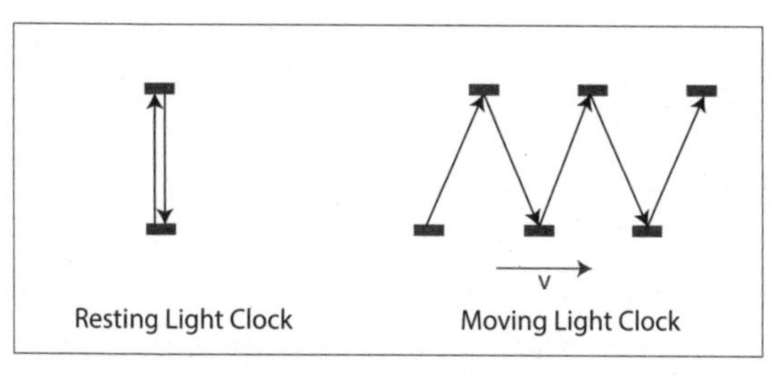

Image 4: *The light clock from the perspective of the resting (left) and moving (right) spectators.*

protected zone. In Berlin he can withdraw at will to Haber's institute and mull over his thoughts as he walks in the Grunewald and sails on the Wannsee. He is free to enjoy the serenity of the new library building on Unter den Linden, have discussions with Academy members, or consult researchers from the Physical Society.

However, the desired peace is elusive. Berlin is not the "Isle of the Blessed" Aristotle talked about, where all that remains is only thinking and contemplation. Quite the contrary. Shortly after his arrival in Germany, a storm sweeps through Einstein's life—two concurrent catastrophes, of which the private one was foreseeable, while the public one will remain forever inconceivable.

Part II: The Battlefield

"We do not eat each other,
we just butcher one another."[115]

— George Christoph Lichtenberg

4. Ultimatum

Night Train to Zurich

Nine o'clock in the evening, Wednesday, July 29, 1914. A time of homecoming. With their bags and suitcases, people who had suddenly cut short their vacations flood the Berlin Anhalter train station. Following the Kaiser's example, they have returned to Berlin ahead of time. They step off seemingly tiny trains into a 34-meter-high and 170-meter-long concourse, a palace of iron and glass that emerged at the beginning of the Wilhelmine period. As they stream toward the exits, the echo of their voices dies under the huge vaulted ceiling.

On the departure side of the hall, Albert Einstein is standing on the platform, looking at the long-distance train that has just left Berlin for Zurich. Fritz Haber, who is at his side, will stay with him the entire evening. As the last car leaves the long platform behind, Einstein breaks down in tears, howling "like a little boy."[116] He weeps for his two sons, Hans Albert and Eduard, who are returning with their mother to Switzerland.

It is one of the bitterest hours of Einstein's life. Parting from his two boys is extremely hard for him. In a letter to Elsa at the end of July 1914, he said that he considered the separation from his children to be a great misfortune and that he would have been a monster had he felt otherwise. "I carried around these children countless times day and night, drove them around in strollers, played with them, exercised and joked with them."[117] In another letter to Elsa from the same time, Einstein wrote that he could act in no other way, despite

Image 5: *Mileva Einstein with the two sons Eduard (left) and Hans Albert (right) in 1914*

the children, and that he needed time to recover. "Such a thing is somewhat akin to murder."[118]

Right up until they parted, little "Tete" still cheered with delight at the sight of his father. Having just turned four years old, he could not grasp the situation between his parents. Now he was sitting next to his ten-year-old brother in the departing train.

Mileva left Berlin holding a deep grudge against her husband. She had made every effort to prevent the separation. Why had he taken her along to Berlin? To uproot her only to give her marching orders? Clara Haber, with whom she stayed, had also tried to mediate between the spouses. But Albert had reached the point of no return. He pushed Mileva toward divorce with unyielding harshness. In the end, her last spark of hope was smothered by his selfishness and emotional coldness.

Moving to Berlin had made Mileva's situation hopeless. These last few weeks had weighed on her like a nightmare. Instead of settling in their new environment, she hardly set eyes on her husband. Albert disappeared sometimes for days, without notice. She felt lonely, rejected, and betrayed. Under different circumstances she might have packed her suitcases earlier, taken the next train, and together with the children sought refuge with her Serbian family. But a journey to Vojvodina was unthinkable in the summer of 1914.

Since June 28, when a Serbian assassin fatally shot the Austrian successor, Prince Franz Ferdinand, and his wife, Sophie, in Sarajevo, everybody in Europe had been counting on the Austro-Hungarian monarchy to order a military strike against Serbia. The Austrian rulers assumed that the instigators of the atrocity were to be found in Belgrade. Presumably the assassin belonged to a terrorist network, the Black Hand, who worked toward integrating all the Serb-inhabited territories, including Vojvodina and other southern regions of the multinational Habsburg state, into a greater Serbian empire.

In the nineteenth century the Bosnian capital, Sarajevo, was still part of the Ottoman Empire. It was in 1908, in the run-up to the two Balkan wars, that Austria-Hungary first annexed the Bosnia-Herzegovina province, in which almost half of the population was Serb. Since then, Franz Ferdinand had been toying with the idea of turning the dual monarchy into a tripartite state, Austria-Hungary-Croatia, namely to establish a monarchy in the south under Serbian control as part of the Habsburg monarchy. This plan could not be reconciled with the dream of a great Serbian state.

As expected, after the assassination of the successor to the throne and his wife, the press talked of the "murder of the sovereign by Great Serbian nationalists." The government in Vienna immediately announced measures to "contain the Great Serbian agitation in the south of the Monarchy."[119] The Habsburg Empire armed itself for a war against the neighboring state, thereby cutting off Mileva from her closest family.

Since leaving Zurich she had been entirely on her own. Her husband enjoyed the company of his lover, his mother's Swabian cooking, and the ambience of a city of science. In contrast, her own life constantly narrowed to fewer people and events. Others now occupied a place at Albert's side. His circle of acquaintances grew from day to day.

Against all expectations, Albert had successfully settled in Berlin, as he let his friends and colleagues in Switzerland know. Only the clothing drills he had to undergo at a few uncles' command, "in order not to be associated with the scum of local humanity," disrupted his peace of mind.[120]

Einstein had begun his career at some distance from academic research but in active exchange with his wife, Mileva, and the members of "Olympia," his private academy. This circle usually met in the apartment of their president, "Albert, Knight of the Steissbein" (*Steissbein*, which in German rhymes with *Einstein*, literally means "tailbone" but is used in the more humorous sense of "backside"). There, in a small group surrounded by tobacco smoke, they would read philosophical works by Baruch de Spinoza or David Hume and discuss methodical issues in contemporary research. Now he was a member of the time-honored Prussian Academy of Sciences where, being in his mid-thirties, he felt rather out of place. The aged gentlemen around him, who addressed each other with "Excellency," fell regularly into dignified sleep during the academic meetings.

The university colloquium in the large auditorium at the Physics Institute at the Reichstagsufer, a place for thinking where the participants stood out more for their scientific curiosity and less for their bombastic grandeur, was much livelier. Such a cluster of excellent physicists was to be found nowhere else in the world, Einstein later noted.[121] The high-caliber discussion circle offered him the opportunity of exchange with experienced researchers and promising young talents in a comparatively casual atmosphere. "It is tremendously stimulating here," he wrote to his friend Paul Ehrenfest.[122]

Within a few weeks of his arrival in Berlin, Einstein was elected to the German Physical Society. Here too, all eyes were on the newcomer. Einstein, in the words of the contemporary writer Robert Musil, was eager to deny the existence of space and time. "But not dreamily from a distance, as the philosophers sometimes do—which everyone then instantly excuses by blaming it on their profession—but with reasons that emerged suddenly like a car appearing out of nowhere and that were terribly credible."[123]

Max Planck welcomed the special theory of relativity, because it formed a common conceptual foundation for natural phenomena in mechanics, optics, and electrodynamics. It came very close to his ideal of a unification of physics. But not all scholars agreed with his estimation, that with Einstein they had a new Copernicus in their ranks. Even for a seasoned physicist, it was not easy in the big city, where synchronized clocks ticked simultaneously, to dive into Einstein's new theory of time and a world where bodies moved toward or away from one another with almost the speed of light. In 1914 the German Physical Society published a whole series of articles critical of the theory of relativity.[124]

Relativity Put to the Test

Einstein had abandoned the thought of universal cosmic time in 1905. Through the example of the two bolts of lightning striking a train, he made clear why the idea of an absolute simultaneity is misleading once one approaches speeds that are similar to that of light. The two bolts of lightning at the back and front of the train seem simultaneous to an observer standing on the platform, but not to a train passenger who is moving very fast in relation to the standing observer.

This relativization of simultaneity means that observers who are moving against each other perceive time intervals differently. The faster they move in relation to each other, the more the rate of their "clocks" will deviate from each other. Time does not run equally for both. According to the special theory of relativity, time in a moving system runs more slowly than in a stationary one.

At the time, Einstein had celebrated the breakthrough toward a new concept of space and time with Mileva, as revealed in a postcard he sent to a member of the Olympia academy: "Totally smashed, both under the table. Your poor Steissbein (Backside) & wife."[125] That was nine years ago. Mileva no longer took part in scientific discussions.

Her husband believed her to be distrustful and unfriendly toward other people. Unfamiliar faces were all she saw in Berlin.

Meanwhile, everything was going well for Einstein in the German capital. Although his first lectures there drew some disgruntled criticism, he had a thick skin and was delighted with the attention his theory received. The effects of relativity, such as the dilation of time, were so spectacular that the Institute for Physics immediately dedicated more sessions to discussing the theory in the spring of 1914.[126]

The sharpest objection was raised by Ernst Gehrke of the Imperial Physical Technical Institute. He had read Einstein's article in the *Vossische Zeitung* and went on to collect and comment on all the newspaper articles about Einstein. From the perspective of the experimental physicist, science thrives "in the field of exact experiment and its logical interpretation."[127] Reliable research is built on experimental methods and clear concepts of classical physics. Gehrke refused to abandon these foundations. Most certainly not in favor of an abstract theory, whose mathematically clad results turned reality on its head.

Gehrke had been criticizing Einstein's theory of time from within the German Physical Society since 1911, as well as in professional journals. He had no doubt about the absolute simultaneity of two events. Rather his criticism concerned the general experience that the flow of time could neither be stopped nor turned back, but, as Isaac Newton wrote, it flows steadily and without relation to anything external.

In Gehrke's opinion Einstein was too hasty in abandoning absolute simultaneity, because he believed that the light speed in a vacuum was constant in all circumstances. But why, of all things, should the speed of light be independent of the motion of the observer and the motion of the light source? What if one day it would show itself as variable?

Although at the time experiments provided no clue to a variable speed of light, this was quite possible. Gehrke and his correspondents wondered about Einstein's "naivety" regarding this issue. His argument could merely imply that "light rays are completely inadequate for determining synchrony in clocks."[128]

Einstein trusted his intuition. The thesis on the constancy of the speed of light could be best integrated into Maxwell's theory of electrodynamics. It was based on very little, but from his point of view these were meaningful experimental findings. In view of the far-reaching conclusions that Einstein drew from the alleged constancy of the speed of light, Gehrke, the experimentalist, considered these tests insufficient. He felt the entire theory of relativity was hanging by a thread.

While Gehrke was expressing his criticism, astronomers were coming up with new measurements. In June 1914, Paul Ehrenfest, who was teaching in Holland, visited Einstein for a week. Except for sightseeing walks, the two scientists used every minute to chat about quantum and the physics of relativity. The question, whether the speed of light is constant or not, did not leave Ehrenfest in peace, as Einstein admitted frankly: "If the speed of light depends in the slightest on the speed of the light source, then my entire theory of relativity, including the theory of gravity, is false."[129]

At his inaugural lecture in Leiden, Holland, in 1912, Ehrenfest had encouraged his colleagues from testing Einstein's thesis. Finally, the Leiden astronomer Willem de Sitter figured out what such a test could look like. De Sitter watched stars that rotated around each other. Although such double stars cannot be identified with the naked eye, they abound in our Milky Way and are easily traceable with a telescope. When one of the two stars is much heavier than its companion, it can be considered as the static center around which the second star orbits. In this manner, two light sources—one static and one mobile—can be distinguished from each other.

This is a matter of a perspective. From one perspective, as though looking down at the stars from above, astronomers see one star orbiting the other. However, in a few cases the double star system is arranged so that astronomers have a lateral view and see the stars from the side. It then appears as though the lighter star moves toward the astronomical observer and then away from him again during its rotation. Consequently, if the speed of light depends on the state of motion of the light source, the receding semi-circular segment of the rotating star would have to seem, from an astronomer's perspective, to be traveling more slowly than the approaching segment. However, de Sitter could determine no such effect.

To Einstein's great excitement, the speed of light remained constant within the scope of accuracy of the astronomical measurements. From 1914 onward, he repeatedly referred to de Sitter's observations in order to take the wind out of his critics' sails. Gehrke remained skeptical. He disputed de Sitter's measurements. Only in the 1970s would these results be impressively confirmed, namely by the accurate measurement of double stars that emit X-rays.

The Twin Paradox

Intense discussions flared up around Einstein's new concept of time, specifically about the question whether processes in a moving system occur more slowly than in a static system—from clock vibrations, to a cigarette's glow, to the human heartbeat. For Gehrke, this meant the descent of science into "chaos." He distrusted the theory of relativity with its claim of universal explanation but could not refute it by means of concrete measurements. Hence, he too had to rely on thought experiments. In order to uncover the entire contradictory nature of the theory, he considered the following case, the still famous "twin paradox."

Two clocks stand next to each other. "One clock repeatedly moves away from the other and then moves back, until finally the two clocks are at rest again next to each other." Now, we first take clock A as the static one and B for the mobile one, and then take clock B for static and A for mobile. "These two processes are in relative terms exactly the same." If, however, mobile clocks run more slowly, then in the first case clock B would lag behind clock A, and in the second, clock A would lag behind clock B. Two processes, identical to one another, lead to different final conditions.[130]

That sounded odd. What if instead of a clock, a person, a twin, was sent on a journey and after traveling at near the speed of light returned to his original point of departure next to his twin? If one follows the theory of relativity, the traveling twin would be younger upon his return than the twin who stayed behind. Suddenly, time travel became conceivable. Did Einstein actually want to go so far as to say that people who move rapidly age more slowly than those who are at rest?

This twin paradox was already known to insiders. In a 1911 lecture in Zurich, Einstein himself sent first clocks then living organisms on an imaginary journey and back to their place of origin. "For the moving organism the long journey lasted just an instant, if the motion occurred at near the speed of light." However, during the journey "the corresponding organisms that remained in their place of origin had long been replaced by new generations." This was an undeniable consequence of the special theory of relativity.[131]

Einstein proved here yet again a keen sense for calling into question great theories. Indeed, he was constantly in search of natural order, but there was a provocateur slumbering inside him, waiting for an opportunity to tease. He was differentiated from other physicists of his time mostly by his ability to apply opposite positions to his own theory, and what's more, he searched by way of an inner discourse for just that which perhaps did not fit into the big picture. Gehrke's objections did not alarm him, for he had anticipated them. Just as he would later taunt quantum theory, he argued against his own theory of relativity. Whoever approved of it was welcome to do so, but they would not be able to get around the twin paradox and would have to bite that bullet.

The twin paradox invited researchers and laymen to play mind games. Let us assume that the thirty-five-year-old Albert Einstein fell passionately in love with Elsa's older daughter, Ilse, who had just turned seventeen—not an entirely absurd thought, as we shall see in the last chapter of this book. The large age difference could hinder a love affair. So the cunning physicist leaves Earth in a spaceship and jets off at 80 percent the speed of light. After he has been away for six years he goes back and returns at the age of forty-one to Berlin.

From his perspective he has aged quite normally during his journey. Yet on Earth, according to the theory of gravity, everything has progressed at a much faster pace. Ten years have elapsed here since his departure. Ilse has turned into a twenty-seven-year-old woman, and if Albert is lucky, she is not yet married and perhaps even delighted with his extraterrestrial proof of love.

One can imagine how Gehrke and other scholars sought to ridicule the special theory of relativity in light of such examples. For "then every train driver has to live longer than his colleague, the postman, who goes by foot only," was their derisive prophesy. If Einstein was right, Bismarck might possibly still be alive and would protect the homeland from downfall.[132]

This was naturally nonsense. Only at extreme high relative velocities, far from everyday technology at the time, does Einstein's special theory of relativity differentiate itself significantly from classical physics. Still, can the theory explain conclusively why in the above example Albert has aged by six years but Ilse by ten?

Consider the journey from both points of view. After Ilse and Albert are separated, each sees the other's clock, according to the theory of relativity, running more slowly by 60 percent. The situation is completely symmetrical. When an hour had passed by on Albert's on-board clock, Ilse's clock on earth shows just 36 minutes—and vice versa. They simply have different time standards. Why is this symmetry broken? How can one of them age faster than the other, in the event that they get together again?

This can be best understood if the two of them compare their watches along the way. To do that they would have to exchange information.

Suppose that Ilse and Albert sent each other a time signal whenever a year had passed from their perspective. This temporal information—for instance, a light signal—would cover the growing distance between Earth and the spaceship. The time this signal would need to cross this distance would always grow longer after Albert's departure. This "Doppler effect"* would reinforce the impression that each other's clock is getting slower.

A simple calculation demonstrates that if Albert were to send Ilse a time signal once a year according to his time calculation, Ilse would receive the first time signal only after three Earth years, the second after six, and the third after nine years. Conversely, Einstein too would receive the first signal from Ilse only once three years had passed according to his clock. At this juncture he finds himself at the turning point of his journey. At this point the situation changes fundamentally. By reversing the direction of his motion, Albert changes his spatial and temporal frame of reference. With that, his judgment of simultaneous events changes—with corresponding consequences, as we will see presently.

Let us look at his return journey from both perspectives. While his super-fast spaceship moves toward Earth, the timing between the signals that he and Ilse exchange annually is constantly becoming shorter. The annual messages reach each other faster. During the entire return journey, this effect overlaps with relativistic time dilation. Since the effect is numerically greater, Albert and Ilse see the other's clock running not more slowly but somewhat faster.

Up to Albert's turning point, Ilse received a signal from him three times, the last one only after nine Earth years. This changes during his return journey. Within one Earth year she receives three further annual messages—and then Albert is there! Altogether ten Earth years have passed by the time of his arrival. But given her previous exchange of information, Ilse knows that six years must have passed for Albert, since altogether she received six time signals from him.

* The effect produced by a moving source of waves in which there is an apparent upward shift in frequency for the observer toward whom the wave is moving, and an apparent downward shift in frequency for the observer from whom the source is receding.

From Albert's point of view, each of the journeys, there and back, lasted three years. True, only one single message arrived from Ilse on his journey in space. But during the three-year return journey, he too was contacted with a time signal that arrived on board his spaceship every four months. He concluded from the total of ten messages that ten years must have elapsed on Earth.

Contrary to Gehrke's assumption, the theory of relativity is not self-contradictory at this point. The information exchange between Earth and the spaceship sheds new light on the seemingly paradoxical situation. The amount of time as measured by Albert on his outward journey (three years) and return journey (another three years) is shorter than the amount of time as registered by Ilse (nine years and one year). This asymmetry came about as Albert turned around, his spaceship accelerated, and he left his initial frame of reference.

Einstein answered Gehrke's objections along this line in a physics colloquium: "Clock B, which was moved, is slow, because in contrast to clock A, it suffered accelerations." The acceleration phases by themselves are insignificant for the amount of time difference between the two clocks. "However, their presence requires that clock B would run more slowly than clock A."[133]

While he was defending his theories, Einstein saw no possibility of actually using clocks to measure the predicted path differences. For at the relatively low speeds that could be achieved by the airplanes and trains of his day, the expected time dilation was immeasurably small. And since even in the future the speed of such vehicles would be by order of magnitude less than the speed of light, Einstein doubted that anyone would ever be able to perceive the odd time delay.

Decades later, enormous progress in the technology of clocks would enable the measurement of Einstein's predicted deviation from classical physics with the accuracy of moving clocks. In October 1971 the American physicists Joseph C. Hafele and Robert E. Keating took four cesium beam atomic clocks aboard a commercial airliner. They had to purchase an extra flight ticket for the bulky instruments with the name "Mr. Clock" on it. With this equipment they orbited Earth, first eastward and in the following week westward. Before the flight they synchronized the clocks with structurally similar atomic clocks in the United States Naval Observatory in Washington, DC. After the flight they re-examined the clocks. The difference between the measured values was only a billionth of a second, but it conformed entirely with the theory of relativity.

Such tests were not even remotely imaginable at the beginning of the twentieth century. This did not detract from Einstein's trust in his theory, not even when his adversaries were undeterred by logical arguments and a few formulas. Almost too bored, he wrote to a former Swiss colleague in June 1914, saying that he had just spoken to Gehrke. "If he had as much intelligence as self-confidence, it would have been pleasant to have a discussion with him." In the same letter he announced he would soon take the next hurdle and speak at a colloquium about his general theory of relativity, which elicited "high esteem as much as disbelief" from his colleagues.[134]

Curved Thoughts

One of the disbelievers was Max Planck, who very much admired Einstein's special theory of relativity. He expressed his skepticism in front of a large audience, on, of all days, July 2, 1914, the date of Einstein's ceremonial admission into the Prussian Academy of Sciences.

Despite the heavy heat on this summer day, many guests arrived at the library on Unter den Linden. Under the hall's gilded ceiling they listened to the sometimes abstruse speeches of the Academy members and marveled at the "gallery of wonderful strong minds."[135] After Einstein expressed his gratitude for being able to pursue his scientific studies free of the troubles and worries of a practical profession, Planck welcomed him as a scholar whose "true love belongs to that approach to work in which the personality unfolds most freely, in which the imagination plays its wildest game." Planck felt that Einstein was thus also at risk of occasionally losing himself in an all too dark region.[136]

In Planck's view, the general theory of relativity was a "dark region." And he could not withstand the temptation to announce his objection: For Einstein, the principle of relativity in its current form was not satisfactory because of its preference for constant movement. But was not exactly this preference an important and valuable characteristic of the theory? "The laws of nature we are looking for continually present certain restrictions, namely a special selection from the infinitely diversified range of any conceivable logically consistent relation." Perhaps one could associate the preference for constant movement with the special privilege that distinguishes the straight line from all other spatial lines.[137]

Planck illustrated his reservation with a geometrical metaphor: the straight line should have priority over the bent. Einstein doubted this in particular. His general theory of relativity is about curved space, which we will discuss in detail later. In it the shortest connection between two points is no longer inevitably straight. His studies showed, among other things, that a light ray that passes near the sun will be bent. It would be possible to measure this deflection of light during a solar eclipse.

This was a courageous forecast, for the next solar eclipse was imminent. The Prussian Academy provided the funds for observing this eclipse on August 21, 1914, from southern Russia. The research expedition, organized by the astronomer Erwin Freundlich, was expected to lead to clarification.

Did Planck intend to anticipate the results of the journey with his remarks? Did he want to say that Einstein had curved thoughts? "Whatever the result will be," his speech concluded, "in any case we are facing a valuable enrichment of our science, where, as we may say with pride, it is easier than in other fields of science to resolve the sharpest factual contrasts by personal esteem and cordial disposition."[138]

Einstein had no reason to doubt Planck's friendly disposition. But he perceived a deep gap between the expectation Planck and other physicists had regarding his contribution to quantum physics, and their disbelief in relation to his general theory of relativity. Einstein stood alone in Berlin in his view of gravity. It did not help that he wrote to Planck a few days after the public session in order to explain his views one more time.[139]

Since moving to Berlin, Einstein had hardly been able to work on the general theory of relativity. There was always so much going on in the big city that he hadn't even had time for music. Instead of playing the violin and riding his favorite mathematical-physical hobby horse, he took an active part in the scientific life of Berlin and occupied himself more than ever with quantum theory, which likewise revised basic concepts of physics.

In this regard, meeting with young researchers such as James Franck and Gustav Hertz was particularly inspiring. Thanks to the financial support of the Solvay Foundation, they had built a first-class laboratory in Berlin where they produced electron–atom collisions. The physicists showed that atoms always absorb or give off the same amount of energy, namely the same number of quanta units.[140] Einstein was fascinated by the "striking confirmation" of the quantum hypothesis for which the two researchers would later receive the Nobel Prize.

In Berlin Einstein became increasingly more involved in debates about the structure of matter and around Bohr's atomic model, which had recently been made known. He struggled with quanta and intended to speak on quantum theory before the German Physical Society. The title of the lecture, planned for July 24, was a reference to his two Berliner supporters: "A theoretical perspective on the thermodynamic derivation of Planck's radiation formula and on Nernst's heat theorem."[141]

Einstein slaved away on his lecture despite an almost unendurable heatwave that summer. The temperature in July 1914 was considerably above 30 degrees Celsius, and many city dwellers made a morning pilgrimage to the Wannsee or the Müggelsee or traveled to the beaches of the Baltic Sea. Einstein, however, continued to go to his office at the Kaiser Wilhelm Institute for Physical Chemistry and Electrochemistry after scribbling mathematical exercises nonstop on wrapping paper during breakfast.

Saber Rattling

In the meantime, the Kaiser had taken to the sea as he did every summer with his yacht, the *Hohenzollern,* and was sailing along the Norwegian shoreline. Just before leaving Berlin, Wilhelm II adopted a highly risky political position. Contrary to his early Balkan policy, after the Sarajevo assassination he gave Austria a free hand to take military action against Serbia, sooner rather than later. Should Russia attack, the German Reich would "comply with its commitments to its allies and its old friends and…remain loyal to Austria-Hungary."[142]

This "carte blanche" was based on a misjudgment of the political situation. Just prior to his departure from Berlin, the Kaiser confirmed that he did not believe there would be any major complications. Like King George V, Czar Nicholas II would not stand at the side of the prince's murderers. The Kaiser expected instead that his two cousins, in England and in Russia, would demonstrate "monarchical solidarity against the regicidal band of robbers."[143]

Kaiser Wilhelm anticipated a quick retaliatory strike of Austria-Hungary against Serbia—without a declaration of war—while public indignation over the royal murder was still strong all over Europe. But this didn't happen. While key officials in Vienna believed there was no alternative to a military intervention, the thought of a preventive strike was abandoned because most army units were still on harvest leave. The soldiers were bringing in the crops and could not be ordered to return to their barracks so quickly.

Furthermore, Hungarian Prime Minister Istvan Tisza refused an immediate invasion of Serbia. He feared for the stability of the multinational state and warned Vienna about incensed southern Slavs—new centers of conflict in the regions of the Habsburg Empire inhabited by Serbs—and about a Russian intervention. His proposal: Vienna should make concrete demands and present Belgrade with an ultimatum, and mobilize against Serbia only if the demand was rejected. But his proposal did not come closer to a peaceful solution to the conflict. With the exception of Tisza, who favored presenting Serbia with severe but acceptable demands, all in the Austro-Hungarian Ministerial Council agreed the ultimatum had to be strongly worded in order to guarantee rejection.[144]

Wilhelm II set off on his northern cruise without knowing what his allies were planning to do. The German Reich Chancellor himself, Theobald von BethmannHollweg, who had advised the Kaiser to take his vacation trip, retreated to the Hohenfinow castle in Mark Brandenburg for his summer vacation. The chief of the general staff, Helmuth von Moltke, and the minister of war, Erich von Falkenhayn, went on holiday as well. Consequently, the public impression in Germany during July 1914 was that the conflict would end up like many former crises on the Balkan Peninsula.

Einstein's correspondence of this time does not indicate how seriously he took this "saber rattling." While others went on holiday, he stayed in the metropolis. On July 7 he even called off the mountain hike that he and Paul Ehrenfest had planned for the end of the month, because he had received alarming news: his mother Pauline was diagnosed with cancer and had to have surgery.[145]

The effect of this news on Einstein's private life cannot be assessed from his sparse letters. In the past three months he had seen his mother more often than he had in a long time. Although Mileva would have liked nothing better than to build a wall between Albert and his relatives, Einstein frequently traveled between Dahlem, Schöneberg, and Charlottenburg. The children were waiting for him at home in Dahlem, his lover in Schöneberg, and not far from there his fifty-six-year-old mother, who until recently had still taken care of him and who now needed his undivided attention.

In view of Pauline Einstein's upcoming lower abdominal surgery, Albert and his Swabian relatives came even closer together. Aunt

Fanny, Uncle Rudolf, Uncle Jakob—all were looking after the patient. Albert himself found the strongest solace and assistance in the company of his cousin Elsa, who more than ever became the woman at his side.

Elsa respected Albert's need for independence. She had come to terms with the fact that solitude was a necessary source of energy for him and that he subordinated almost everything else to his scientific curiosity. But she had expectations of him that coincided with the wishes of her parents and those of her sick Aunt Pauline: she wanted to finally have a clear relationship.

While still in Zurich, Albert had promised Elsa that after his relocation he would live with her and they would run a small "Bohemian household" in Berlin. When Elsa confronted him shortly thereafter about separation from Mileva, he responded by telling her that getting a divorce was anything but easy "when one has no proof for the other's fault." He treated Mileva as an employee whom he nevertheless could not dismiss.[146]

Elsa suffered from this situation—now the family conflict was out in the open. Evidently the Swabian-Jewish clan consciousness increased pressure on Einstein, for suddenly he not only considered separation from Mileva, but also a second marriage. Although he opposed the institution of marriage after the miserable failure of his marriage to Mileva, he nurtured Elsa's hopes. When in the meantime she reproached herself for having destroyed his family, he pacified her by saying that if he had not been so indifferent to the private side of his life, he would have been separated from Mileva long "before I got to know and love you."[147]

A deafening silence prevailed between him and Mileva. Being with her had become intolerable for him. He felt as if he had spent the past few years in prison. He found it inconceivable that she still clung to him. Did she fear the ostracism that a separation involves? Were existential fears preventing her from making a decision? Had she followed him to his new place of activity just because she did not want to put the children through the effects of a separation?

The Tyrannical Spouse

In July 1914, Einstein wanted to know nothing about the needs of his family. He told Elsa he was no longer able to love one woman and be married to another.[148] According to later descriptions by Anna Winteler, the wife of Albert's long-time friend Michele Besso, he moved out of the home in Dahlem that he shared with his wife and threatened to set her up with a tenant in the apartment in order "to virtually push her away."[149]

As Mileva resisted him, he gave her an ultimatum, making heartless conditions for further living together, clearly for no other reason than to bring about the final breakup. Mileva had to: a) take care of his clothes and laundry, serve three proper meals a day *in his room*, and tidy up his bed and workroom; b) renounce all personal relations with him, as far as they were not absolutely required for social reasons. In particular, he would neither stay home with her nor go out or travel with her; c) "You explicitly commit yourself to paying attention to the following points when communicating with me:

1. You can neither expect any affection from me nor reproach me for anything.
2. You have to immediately cease speaking to me upon my request.
3. You have to leave my bed or workroom immediately without contradiction upon my request."[150]

As wretched as these demands by her tyrannical spouse were, in her total desperation Mileva clung to them. She expressed her willingness to accept the conditions. As mediators between the couple, Fritz and Clara Haber handed Mileva's acceptance to Albert, who immediately responded further, that if he returned to the apartment it would merely be in order not to lose the children—only for that reason. A comradely friendship with her was no longer an option. And should she find it impossible to carry on living together on this basis, he would "give in to the necessity of a divorce."[151]

What sort of perfidious game was he playing with her? Mileva decided she could no longer cope with Albert's callousness. At the urging of Clara Haber, she and the children moved into the upper floor of Clara's executive villa. This did not make things easier for Mileva, even though she had a friend in the lady of the house, who was in the same boat as her, likewise living unhappily at the side of a famous researcher.

Clara Haber had a doctoral degree in chemistry. She would have liked to maintain her professional career, but after marrying Fritz and delivering a son she too found herself in the less desirable role of housewife and mother. Even she, the ambitious researcher, did not achieve what the physician Anna Fischer-Dückelmann declared in her bestseller, *The Woman as Home Physician*: that marriage and profession could be integrated.

In her book, which in 1914 reached a circulation of one million copies in the German-speaking world, Fischer-Dückelmann maintained that the woman of the future would be neither destitute nor exploited. "The more developed her abilities are, the better a woman understands how to combine a marriage and a limited number of children with a profession." Women, then, would no longer be entirely consumed by the business of procreation nor become dulled by the yoke of matrimony.[152]

Fischer-Dückelmann promised women new moral laws that would endow them with freedom and rights—and a man with restraint in his pleasure seeking—as well as liberation from prejudice. For the author, divorce was not taboo. "Should the spouses no longer concur with each other, their love lost, they should not stay in an unworthy relationship, for the wife can live just as well without the husband; they should part peacefully or live without marriage as 'their children's managers' for their sake, when maturity brings them the required peace of mind."[153] Only a few women summoned the courage to take such a step. Blatantly disadvantaged by the Civil Law Code, they risked finding themselves without financial security and being ostracized by society.[154]

Clara Haber and Mileva Einstein had probably never considered a divorce seriously, although both lived through severe crises. Even now, as her husband compelled her to dissolve their marriage, everything in Mileva rebelled against it. His thoughtlessness made a temporary separation appear as the only sensible thing. Fritz Haber drafted a corresponding contract, according to which both sons should stay in Zurich with their mother. In addition Mileva imposed the condition that she would never have to give the children to her husband's relatives, and that she should receive from him 5,600 marks a year, just under half of his salary from the Academy.

On July 24, the same day her husband held his highly anticipated lecture on quantum physics at the Reichstagsufer, the couple had a three-hour talk at the Haber Villa during which Albert agreed to the contractual regulations.[155] Even if this meant his sons might become completely alienated from him, he put an end to the relationship. The way to a divorce was now paved, he wrote two days later to Elsa, who had taken her two daughters on a vacation to Bayrischzell on the border with Austria. Now she had proof that he could make sacrifices for her.[156]

Einstein would not, however, be able to deliver on his promise to marry Elsa right away, as a divorce was out of the question for Mileva, who decided to return to Switzerland with the children. She postponed her departure yet again. Did she hope her husband would back down at the last minute so as not to lose the children?

Last Days of Peace

Meanwhile Einstein's cousin Elsa cut short her holiday in the Alps as did many other summer tourists, for Austria-Hungary had begun to mobilize against Serbia. Europe was under the threat of war. On July 23 the Austrian government's envoy presented the carefully prepared ultimatum to the government in Belgrade. The Serbian answer arrived on schedule forty-eight hours later, agreeing to most of the demands. But the government in Belgrade expressly rejected one of the ten demands: it would be a violation of the Serbian state to allow

Austrian officers to participate in prosecuting suspects in the assassination.[157]

The Serbian government did not reject the ultimatum without first consulting the Russian foreign minister, who promised military support in case of war. Russia in turn could count on French assistance. The French minister of state had just visited the Russian Czar in St. Petersburg and reaffirmed assistance in case of war. Only an igniting spark was missing to initiate a chain reaction and trigger all-out war in Europe. As Austria-Hungary's only reliable ally, Germany played a key role.

The German Social Democrats immediately protested against the "frivolous provocation of war." "We want no war," read the posters with which the party called for protest meetings to be held on July 28 across Germany. Not one drop of a German soldier's blood should be sacrificed to the "power seekers of Austrian rulers."[158]

One day before Mileva's departure, workers and employees demonstrated all over Berlin against the looming armed encounter. Meeting places such as the union hall on the Luisenstadt Canal or the Friedrichshain Brewery had to close early due to overcrowding. More than 100,000 people took part in thirty-two anti-war rallies in the Reich's capital.

"Afterward, participants from all the meetings, in large numbers, mostly crowds of thousands, attempted to push toward the city center," reads the chief of police's situation report. It was only by using armed force and making arrests that the moving masses, howling and roaring revolutionary songs, were scattered.[159]

Some demonstrators did get over the barricades. In the middle of the night they roamed the Unter den Linden boulevard, singing. Young German nationalists immediately formed a counter demonstration. To the workers' slogans "We want no war" and "Long Live the international fraternity," these young men responded with "The Guard on the Rheine" and the national anthem, "Hail to Thee in Victor's Crown."

Commenting on the protests, the *Berliner Tageblatt* said, "We do not condone street demonstrations at this time of international conflict. But if young war enthusiasts are allowed to express their views loudly in the streets, those who rightly desire international peace and have a clear idea of the terrible pain of war should be allowed to do the same." The German people, upon whom war would bring so much misery, must expect of those who bear responsibility at this difficult time "that despite resistance, no opportunity to prevent the danger will remain unexploited."[160]

Conservative journals and the Kaiser adopted entirely different tones. The *Preussische Zeitung* (Prussian Newspaper) talked about "high treason": Willhelm II said the anti-military activities of the left-wingers should not be tolerated. In case of recurrence he would have the leader and all of them jointly and severally imprisoned.[161]

After interrupting his trip to Norway, the Kaiser didn't read the Serbian answer to the Austrian ultimatum until Monday, July 28, and assessed it as a submissive compliance. "This is more than anyone could have expected! A big moral success for Vienna. But this rules out any reason for war."[162] He immediately drafted a letter to the allies, reinforcing that there were now no longer grounds for war. Instead he proposed a temporary occupation of Belgrade as a "bargaining chip," in order to hold Serbia to its promises and give the Austrian army "an utterly honorable satisfaction" after having been repeatedly mobilized in vain. He himself wished to act as mediator.[163]

This initiative, which was not expected by the Kaiser's close circle of advisors, is a testimony to the striking information deficit of Wilhelm II, the grandiose gestures typical of his behavior during the July crisis, and the uncoordinated actions of those with political responsibility. Believing that he personally could prevent the war, the Kaiser turned down the suggestion from England to convene an international peace conference. As his document was brought to the foreign office to be forwarded to Vienna, events had already overruled his plans. Austria-Hungary had that morning declared war on Serbia.

If that were not enough, the Reich's Chancellor Bethmann Hollweg and the foreign office in Berlin inexplicably delayed sending a telegram to Vienna, and when they did, no mention was made that in the German Kaiser's view there was no longer any reason for war and that he was offering to act as mediator. The Chancellor's main concern was "that the responsibility for the potentially spreading conflict" would rest with Russia under all circumstances.[164] In case it should indeed come to war against Russia and France, the impression must be given that Germany was forced into war by Russia and that it was a matter of defense. This, in his opinion, was the only way to keep England out of this "mess" and win the support of the German people for the war. His plan was successful, at least in his own country. Over the next few days the German press almost unanimously warned against the "terrible danger of Russian barbarism" and the "oppressive regime" of the Czars.

Czar Nicholas II and Wilhelm II were cousins, who signed their mutually distrustful telegrams to one another "Nicky" and "Willy." "Nicky" was imploring the German Kaiser to restrain Austria-Hungary, saying that a weak state had declared an unworthy war. "Willy" thought a war against Serbia was anything but "unworthy" and cautioned Russia to exercise restraint. "Naturally, military measures on the part of Russia…would wreak havoc, which we both wish to avoid, and endanger my position as mediator, which I have readily assumed following your appeal for my friendship and assistance."[165]

However, the Czar and the Kaiser were no longer masters of the situation. Their own advisers played them against each other. At the same time the military pressure on the two monarchs increased hour by hour, for the action of the general staff followed its own logic.

For the German army, every day counted in case of a war on two fronts. The German generals wanted to send their armies against France and take Paris even before Russia concluded its deployment of several weeks. No alternative was provided to this "Schlieffen Plan" with its extremely short timeframe for deployment. Political scientists, such as Herfried Münkler, estimated that this arrangement was fatal and served to escalate the conflict.[166]

On July 29, the first Austrian shots were fired at Belgrade. Meanwhile the Chancellor of the German Reich sent a telegraph to Saint Petersburg: if Russia carried on with its military preparations, the German Reich would be forced to mobilize. Thereby he further accelerated the crisis. The alarm bells were ringing anyway at the Russian foreign ministry. The Czar was urgently advised to order the general mobilization in order to round up troops from across the vast empire in time, and to prepare for all possibilities.

Driven by mistrust, Czar Nicholas, who was just as fickle as Wilhelm II, immediately agreed, that very same evening. But then, at 8:30 PM Central European Time, shortly after the mobilization order was sent to all parts of the vast empire, Nicholas suddenly revoked the order. A telegram that had just arrived from his German cousin has made him think again—a last, desperate attempt to stop the already moving military machine.[167] To no avail.

At the *Anhalter* train station where Einstein and Haber have just seen off Mileva and the children, newspapers with headlines such as "Declaration of War" are being snatched from the hands of the news vendors. Were Austrian troops already advancing against Belgrade? Was Russia refraining from action? Had the English mediation achieved anything? How would the German Kaiser react?

Just the previous day, Haber had submitted a request for a holiday. He wished to travel to Karlsbad again this summer for six weeks, to cure "gallstones and mind." He wrote, "Should the political conditions develop so that our country becomes embroiled in a military entanglement, I intend to return from my holiday."[168]

In the letters Einstein wrote in the last days of July and first days of August he makes no mention of the tension felt by Haber and those who had cut short their summer vacations to return home. Not once is the word "war" mentioned in them. Consumed by having had to say farewell to his children, he shut everything else out.

Einstein held his boys in his arms one last time on the platform at the train station. Conscious of his fault, he cried over it the whole afternoon. He was afraid that Mileva was going to put up a wall between him and the children, so that "their mental image of their father will be systematically ruined."[169]

Nevertheless, deep down he felt that the separation from her was a "question of survival" for him. He could no longer tolerate his wife. As he confessed to his friend Heinrich Zangger, the fact he did not find the strength to make such a decision earlier was now incomprehensible to him. "Partly because my means did not allow for a separate life."[170]

His longtime friend Michele Besso had come to Berlin for the sole purpose of accompanying Mileva and the children to Zurich. From the years they shared together in Switzerland he knew how hard it was for Mileva by the side of the genius physicist. From now on, Besso and his friend Zangger would take the role of mediator between the estranged couple and at the same time pay particular attention to the welfare of the children. A photo taken in 1914 shows Einstein's two sons with their mother: "Tete" stares at the camera questioningly, Hans Albert, the older, is sulky.

Before the boys climbed onto the train, their father gave them one last kiss. The doors closed, the engine blew its white smoke into the summer evening, and the train headed south through one of the three brick arches at the end of the concourse, one of the last trains to operate on a civilian schedule.

Einstein stayed on the platform with Haber. "I would not have managed that without him."[171] Haber stayed with his distraught colleague until deep into the night. At this point, he was the only one among the Berlin researchers who knew about Einstein's relationship with Elsa. Haber advised him not to appear alone with her in public in the near future, so as not to provoke gossip, and offered to inform Max Planck before he or any of their other close colleagues heard rumors of the affair.

The next day, in an attempt to regain his composure, Einstein attended a lecture at the Prussian Academy "On the Energy Turnover in Photochemical Processes." First he visited his sick mother to deliver the news; Elsa's parents already knew. The family felt the alimony he agreed to pay Mileva was too high, but they were happy about the separation and his intentions of marriage, although halfhearted. Pauline Einstein reacted effusively: "Ach, if only our poor Papa had lived to see this day!"[172]

5. "Europe, in her insanity, has started something unbelievable."

War According to Schedule

"Our highly acclaimed technological progress, our civilization in general, is like an ax in the hands of a pathological criminal."[173] When Albert Einstein drew this bitter conclusion, the world had long been at war, and all the technological innovations, from cars to airplanes right up to telephones and earplugs, had become part of a gigantic war machine that increased death and suffering by unimaginable degrees. Longtime hopes for progress had turned into destruction.

World War I started with the railroads. They epitomized "the nineteenth-century industrial capability of societies," concluded historian Jörn Leonhard, and symbolized the aspiration for technological progress, mobility, and speed.[174]

In August 1914, the massed armies advanced with breathtaking precision to the front lines. In the German Reich the entire railway transportation system was readjusted on the first day of mobilization.[175] All the freight trains were unloaded and modified for the transportation of soldiers and horses, vehicles, and heavy guns.[176] While the gathered reserves and volunteers were transported to the barracks, the railway provided special trains in order to bring the last summer vacationers back to their hometowns.

On the night of the second day of mobilization, the military timetable went into full force. From that point forward, standardized transport trains rolled on unceasingly through the country at the same average speed, tailored to the performance ability of the robust Prussian freight train engine, the G7. In the first weeks of August, 20,800 mobilization trains as well as 11,100 deployment trains were set in motion, bringing around 3.1 million men and 860,000 horses to the front.[177]

In contrast to the German–French war of 1870–71, the military used scientific knowledge, which in itself appeared harmless, such as astronomical and instrument-based time measurement, to further accelerate the transports. In his last speech to the Reichstag, on March 16, 1891, Field Marshal Helmuth von Moltke spoke about the urgent need for a uniform time measurement throughout Germany in order to successfully manage the railway system and coordinate a mobilization in times of military exigencies.[178]

In the summer of 1914 the Field Marshal's nephew, Helmuth von Moltke the Younger, launched the mobilization, meticulously planned according to a uniform time. After the railway administration was thoroughly trained as part of the annual Kaiser military maneuver, it proceeded without significant incidents.[179] The bridges over the Rheine groaned under the transport load. Between August 2 and August 18, 2,150 vehicles alone crossed the Hohenzollern Bridge on their way west, one every ten minutes. According to the Schlieffen Plan, the preferred destinations for the troops were the train stations on the Belgian side, in order to surround France on the north and encircle its army within six weeks. While German soldiers were crossing the border into the neutral neighboring country, a European disaster of epic proportions was about to start.

"Europe, in her insanity, has started something unbelievable," Einstein wrote to his colleague Paul Ehrenfest on August 19, the day the German army occupied the Belgian city Louvain, where hundreds of civilians lost their lives and the university library, containing 230,000 books, was set alight. "In such a time one sees to what a sad bestial race man belongs."[180] The next day, German soldiers marched into Brussels, where in the previous year Einstein had attended the Solvay Conference and debated the structure of matter with researchers from around the world. August 20 was also the first day of the lengthy siege of the commercial city Antwerp, home to Albert's Uncle Caesar, to whom he sent his first essay many years ago.

Einstein's investigations were directly affected by the action along the Eastern Front. Just before the beginning of the war, the astronomer Erwin Freundlich and two other colleagues left for Russia, carrying expensive photographic equipment in order to observe the solar eclipse on August 21. The expedition's goal: to test Einstein's prediction that starlight would be deflected by the sun's gravitational field. Einstein waited feverishly for this expedition. He had been corresponding with Freundlich for years about the possible confirmation of his theory of gravity. But instead of experiencing the eclipse, Freundlich was taken as a prisoner of war. Einstein was very concerned about him. "What a horrible picture the world is offering now," he wrote to a friend in Zurich. "Nowhere is there an island of culture where people maintain human feelings."[181]

It was incomprehensible, how things had come to this. Similar to the writer Robert Musil, Einstein held it absolutely impossible "that the big nations that are constantly coming closer through European culture, could let themselves get carried away into a war against each other." Many had opposed the war before it broke out. But once it started, the world was split into Germans and anti-Germans, and the individual's only value was his elementary skill in defending the tribe. "Loyalty, courage, subordination, performing duties, sobriety—this is the range of virtues today that make us strong, ready to fight on the first call."[182]

First and foremost, young merchants and business people, academics and students volunteered for active service.[183] A regulation was issued at the onset of the war regarding "early final examination" so that high school students could rush to the front. Not one student failed the early examinations. In Berlin the composition subject on the final exam was "The Importance of the Railroad in the Present War."[184]

The number of volunteers, though, was no longer as high as reported in the press. While newspapers were printing photos of flag-waving fraternity members, movie theaters were showing propaganda films about the enthusiasm for the war in the country, and writers were putting into words their exhilaration over being one with the fatherland, tears flowing in the train stations as relatives, sons, and fathers headed off to war. Summing up his impressions, a clergyman from the Moabit workers' quarter suggested that the response from the poor reflected their immediate reality. "The actual enthusiasm—I should say, the academic enthusiasm such as can be afforded by the educated, who do not have immediate worries over food—appears to me to be missing."[185] And a journalist from the daily *Berliner Lokal-Anzeiger* wrote that neither the processions in the streets nor the signs of vague fear and nervous anxiety could be claimed as the prevailing mood in the country. Rarely do we find a more classic example of "mixed feelings."[186]

The Only Sober One Among the Ecstatic

Although German by birth, Einstein arrived in Berlin a Swiss citizen. When in 1913 Fritz Haber pointed out that his appointment to the Academy meant he had to become a Prussian citizen, Einstein declined. He had struggled to be accepted as a citizen of the freedom-loving country of Switzerland, which had not engaged in war for centuries and had promoted diversity and openness in place of cultural integration, where people whose mother tongue was German, French, or Italian lived together. Einstein made accepting his call to Berlin conditional upon "making no changes in regard to his citizenship."[187] Nevertheless, he was frequently regarded as German during the war, partly because his German colleagues were glad to call him one of their own, both because of his origin and his outstanding achievements.

The great confusion about Einstein's citizenship began much later, namely with the awarding of the Nobel Prize to him in 1922. At that time, Einstein was on a world tour. A Swiss envoy wanted to accept the prize in Stockholm on his behalf. But the Germans, making every effort to improve their image abroad, vigorously claimed Einstein as a citizen of the German Reich, and despite some missing evidence, they succeeded. The legal debate over whether Einstein was Swiss or German continued until the mid-1920s, due largely to his own rather dubious statements. Finally, Einstein gave way and accepted German citizenship in addition to his Swiss.

During World War I, though, the German authorities determined he was a Swiss citizen. Einstein felt he belonged to an international community of scientists, and suffered from the collapse of European civilization, as did all the Swiss. That set him apart from his close research colleagues, Max Planck, Walther Nernst, and Fritz Haber, but also from industrialists like the physicist and head of the AEG Walther Rathenau and the chemist Carl Duisberg, chairman of the board of the paint factory, Leverkusen, although all of them had cultivated international contacts and had traveled the world.

They too did not long for the war. On the contrary. Disapproving of the chauvinism of some academics, Planck had emphasized time and again that the Prussian Academy's affairs were peaceful.[188] Rathenau and Duisberg viewed a war as untenable for economic reasons. "A question such as whether Austrian commissioners must collaborate in the Serbian investigation into hostile activities is no cause for an international war," Rathenau warned in the *Berliner Tageblatt* on July 31, 1914.[189] And there is much to suggest that industrial researchers like Nernst and Haber felt the same way.

A few hours after war was declared, hundreds of thousands assembled in front of the Berlin Palace, cheering the Kaiser, who suddenly no longer spoke of parties, but only of Germans. Wilhelm II swore to evoke the "nation's unity" for the "defense war." He repeated before the Reichstag that Germany was forced to seize the sword in self-defense and with a clear conscience. "We are not driven by lust for conquest, we are inspired by an iron will, to protect the place God has given us."[190]

The academic elite in Berlin, the center of German power and cultural policy, felt particularly called upon to establish "the nation's unity," namely a determined, ready-for-battle male solidarity. As the rector of a university "that can boast to be famous next to the Hohenzollern Castle not just because of its location," Max Planck perceived this as his duty.[191] "The one thing we know is that we, members of our university, stand together as one man with all our moral feelings and our scientific significance, and will hold out in spite of all the hostile slander, until the truth and German honor are recognized by the entire world."[192]

In early August Planck addressed the students with a speech that adopted the Kaiser's defense rhetoric and blamed the aggression on the enemy alone: after it had demonstrated unparalleled forbearance, the German Reich was left with no choice but to draw a sword against the breeding sites of creeping insidiousness. All the physical and moral forces the nation possesses, gather now and blaze up into a holy flame of wrath.[193] Like most Germans, Planck assumed that "greatness is lying ahead," that as in the year 1870–71, the war would be of short duration and students would be back home by Christmas at the latest.

Walther Rathenau saw through the buildup to the war better than the masses. He did not accept the statements the government were making as reliable facts. "A Serbian ultimatum and a pile of confused dispatches! Had I never seen the backstage of this scenery! Then would I have been able to tolerate the nonsense in the newspapers and sleep." Instead of sleeping, he suppressed his worries in working fiercely: "Nevertheless we must win!"[194]

Rathenau wanted to demonstrate he was indispensable for the German cause and immediately held discussions with the ministry of war. After England had entered the war and threatened to cut off the German Reich from the supply of raw materials with a naval blockade, the industrialist stated convincingly that "our country can probably be provided with materials indispensable to a wartime economy only for a limited number of months."[195] Erich von Falkenhayn, the minister of war, immediately entrusted Rathenau with establishing a raw materials division. It would be decisive for the further course of the war, which lasted not months but years.

After rushing immediately to the barracks in order to volunteer at the outbreak of the war in August 1914, Fritz Haber offered his services to Rathenau. It was now the duty of the German people to "use all our strength to bring down our opponents and bring such peace that will make a similar war impossible for generations to come."[196] The chemist did not find it difficult to emphasize the military relevance of his field. From this time on, he took on the significant role of mediator between the ministry of war and big industry, and from the first months of the war led military researchers at his institute.

Meanwhile, fifty-year-old Walther Nernst marched up and down in front of his house and studied drills and military salutations. Nernst who in his youth was found unfit for military service, had nevertheless joined the Imperial Volunteer Automobile Corps on August 11. He wanted to follow his two sons, who had already been conscripted, to the front. Shortly before he left on August 21, he procured rubber stoppers to use to seal bullets holes in his gasoline tank, in case of emergency.[197]

Nernst had not yet arrived in Belgium when his older son, Rudolf, was killed on the Western Front. Although personally devastated, just a few weeks later Nernst began to develop chemical warfare agents with Carl Duisberg. Haber, who had long been one of Nernst's keenest competitors, would follow his example.

Einstein thanked his lucky stars for being spared from conscription, which forced others into this senseless butchery. As a Swiss citizen he could only consider the war a fratricide.[198] Nationalism was to him a malignant illness that attacked and disfigured people.

Einstein had access to practices that shielded him from a drastic withdrawal of European values. Suddenly he was drawn to Early

Christianity and by the end of August wrote to a friend in Zurich that he felt "as intensely as ever before, it is better to be the anvil than the hammer." Any form of pacifism was already being defamed in the German Reich as weakness and cowardice, as irresponsible as the refusal to follow orders. While others joined in with the triumphant shouting, he was very glad "not to be swept up in nationalistic fervor, thanks to his anachronistic feelings [of pacifism]."[199]

Einstein's Swiss citizenship gave him the right not to take part in this war. At the same time, he felt ashamed of the nationalism of his German friends and colleagues. He had to hide his feelings in view of the excessive war euphoria. Consequently, he observed all the more attentively those who, by dint of their military education and under the pressure of public opinion, rushed into battle, arming themselves with rhetoric. All of a sudden their whole language changed. The war showed unimaginable hatred and pseudo-heroism. It was beyond him, the extent to which scholars, of all people, identified with a wholly militarized state, which posed a threat for millions of people throughout Europe.

In this respect, Einstein felt the same as Hellmut von Gerlach, the editor in chief of the *Welt am Montag* (The World on Monday)—"like a sober person in an all too cheerful company"—in the first days of war in Berlin. Former friends had become fanatics who suddenly believed every word of the imperial government and its generals, which until a few weeks ago were sharply criticized. They had turned into Franco- and Anglophobes.[200] "Away with 'menu,' 'diner,' and 'souper*!" read the *Deutsche Hochschulstimmen* (German University Voice). "Now is the hour to free our language from the yoke of foreign control."[201] Sunlight soap became Sunlicht soap, and the Café Piccadily on Potsdamer Platz was immediately renamed Kaffee Fatherland.

The first news of victory from the Western Front ramped up the mood so much that Carl Duisberg, a visitor in Berlin, became un-

* French for supper.

comfortable. "One speaks…of nothing else but of dividing Belgium, France, and Russia," complained the industry leader. Jingoism predominated, and there was a lack of respectable solemnity such as was found everywhere closer to the Western Front and in the face of the numerous incoming wounded. Duisberg doubted whether the exhausted German troops that were on their way to Paris would be able to fly their flag in victory.[202]

Scholars' War

For the time being the Schlieffen Plan seems to be working. No Belgian stronghold could withstand the hundreds of kilograms of heavy explosives fired from the new types of mortars built by the Krupp factory. "Big Bertha," a 42–cm-caliber howitzer so called for its size, was celebrated in the Great Headquarters as a "trump card that cannot be overtrumped."[203]

In the first weeks of the war, thousands of Belgian civilians fell victim to the German army's reckless advance. The international press reported in great detail the actions of German soldiers against presumed partisans and the pillaging of historical sites. That the magnificent university library in Louvain, with manuscripts dating back to the late Middle Ages, lay now in ashes, was sharply condemned by international public opinion as it would later condemn the destruction of the Reims cathedral and other cultural monuments.

The French philosopher Henri Bergson, president of the French Academy of Political and Moral Sciences, explained that the war that had started against Germany was actually the war of civilization against barbarism. The brutality and cynicism exhibited by Germany, its contempt for any justice and truth were "a relapse into the state of wilderness."[204] Bergson found a sympathetic ear across borders. French intellectuals had been considered the guards of society ever since Émile Zola started a press campaign in France with his open letter "J'accuse!" and obtained the acquittal of the Jewish officer

Alfred Dreyfus. Now they were organizing an international propaganda campaign and demanded that their German colleagues abandon militarism.

German scholars struggled with rejecting these accusations. They had not spoken up in the past with any noteworthy criticism of antisemitism or the judiciary. Any resistance from among their ranks came mainly when their own artistic and scientific freedom was at stake. For example, on March 29, 1914, the day Einstein arrived in the Reich's capital, around 1,200 prominent personalities in art and science assembled at a mass rally of the Berliner Goethe Society. Together, they protested against a bill that represented a new attack on the free development of artistic and scientific life.

Half a year later the Berliner Goethe Society became the incubator of German war propaganda.[205] The president of the society, the writer Ludwig Fulda, wanted to convince academics in the neutral countries, especially in the United States, of the untenability of the accusations of barbarism and at the same time, stop all further attempts to drive a wedge between German academics and the military. Fulda raised his voice against "the lies and calumnies with which our enemies are striving to stain the honor of Germany in the grave struggle for existence that has been forced upon the country."[206] His written appeal, "To the Civilized World" (Manifesto of the Ninety-three), was signed by ninety-three leading German writers, artists, and scientists and printed in all the major newspapers.

"It is not true that Germany is guilty of causing this war," it states. What is more, Germany had done its utmost to avert the war. It is not true that Germany maliciously violated Belgian neutrality. It is not true that the life or property of even a single Belgian citizen had been infringed upon by German soldiers unless deemed necessary for self-defense. "It is not true that our troops have brutally wreaked havoc in Louvain. Having been treacherously attacked by the raging population, they bombarded a sector of the city with heavy hearts as they were forced to retaliate." It is not true that German soldiers disregarded international law. "Those who are allying now with the

Russians and the Serbs, who are presenting the world with the disgraceful spectacle of inciting the Mongolians and Negroes against the white race, have the least right to portray themselves as the defenders of European civilization."

After disputing Germany's guilt in inciting war and the atrocities of its army in Belgium, the manifesto turns to the accusations of militarism, stating that militarism was certainly not in variance with German culture. On the contrary, without German militarism, German culture would have long been wiped off the face of the earth. It was for its protection that militarism came out of German culture "in a land that for centuries has been plagued by raiding as no other. The German army and the German people are one."

The manifesto "To the Cultural World" ends with an ardent appeal: "Believe us! Believe that we will fight this war to the end as a cultural people, to whom the legacies of Goethe, Beethoven, and Kant are just as sacred as its hearth and land! This we pledge to you with our names and honor!"[207]

No public declaration in World War I harmed the reputation of German culture and science abroad as this one did, and none grieved Einstein as much. His disappointment that Haber, Planck, and Nernst lent their names to such a propagandistic and racist appeal, was immeasurable. But not only they. In addition to writers, artists, and scholars of the humanities, the *crème de la crème* of German natural scientists also signed the manifesto, such as Nobel Prize laureates Wilhelm Roentgen, Philipp Lenard, and Wilhelm Wien (all physicists), Emil Fischer, Wilhelm Ostwald, and Adolf von Baeyer (chemists), and Emil von Behring and Paul Ehrlich (medical scientists).

The legitimation of militarism, the relegation of German culpability to invisibility, and the declaration, six times, that "It is not true," without producing any evidence, all trigger international outrage. Fritz Haber, who had personally sent some copies abroad, immediately experienced a negative effect in the USA.[208] The academic world there considered the attempt at justification with pure cynicism. German science and education were now being called into question.[209]

One of the consequences was a massive boycott of the German language in almost all fields of science. In the USA, the number of high school students who chose German as a foreign language declined in the first year of war from 284,000 (24.4 percent) to only 14,000 (0.65 percent), where it stayed until 1922.[210]

Einstein scolded the scholars who, at the outbreak of war, "behave as though their cerebrum was amputated." But he did not criticize the indoctrination of his German colleagues only. Now that French, English, and German intellectuals were vying with one another for war rhetoric, he saw a collective insanity presiding all over Europe, for academics everywhere were convinced of their just cause. Nationalism had "shackled otherwise capable, thinking, sentient men like an epidemic disease."[211]

Nationalistic cheerleading had devastating implications for science. Since the first Nobel Prize ceremony in 1901, more international congresses had taken place than during the entire nineteenth century.[212] With the outbreak of the war many international networks disappeared virtually overnight. The Nobel laureate Wilhelm Wien, a physicist who just a year earlier had a vibrant debate with British colleagues at the Solvay Conference, felt by October 1914 that he would not be able to renew his personal relations with researchers from England. "Even our jovial 'Solvay family' is forever scattered."[213]

"We find sympathies nowhere," stated another signer of the appeal, the Leipzig historian Karl Lamprecht, in the fall of 1914. "The attacks of our enemies were followed by highly clumsy defense with almost unnerving success, and the professors ended up holding the bag in this justification contest. It's not only that nothing was achieved, but that much was spoiled," he confessed in a lecture at the Berlin Urania Scientific Society, whose program at that time offered lectures with titles such as "Our Just Cause" or "We Barbarians" that spoke to the unswerving and obstinate position of the representatives of German science.[214]

Withdrawing to the Ivory Tower

Einstein wrote to a friend in Holland that the international catastrophe depressed him as an international person. "Experiencing this 'great time,' it is difficult to understand that one belongs to this mad, degenerate species that accredits itself with free will."[215] To limit the damage to science, he strove to maintain his own contacts with researchers abroad and prevent the intellectual frontlines from becoming more rigid. The connection to Paris, though, where he was going to present a lecture at the Collége de France in the fall, was cut off. It was only through friends in Switzerland that Einstein learned that Marie Curie was now building a mobile roentgen unit in order to use X-rays for injured soldiers at the front.

Einstein's commitment to international cooperation gained a political dimension from this point forward. More than once in the international arena he protected his German colleagues. In the summer of 1915 he wrote to a friend in Holland that as scientists, their focus was strictly international, whereas one found, to a large extent, chauvinists and hotheads among historians and philologists. All the level-headed people in Berlin had come to regret the famous and notorious "To the Civilized World" appeal. "The signatures were given carelessly, partly without reading the text. Such was the case, for example, with Planck and Fischer."[216]

Einstein's statement was right in that the natural scientists contributed comparatively little to the war propaganda. One can question whether their "strictly international" orientation set them apart from other academics. Weren't Planck and the chemists Emil Fischer, Nernst, Haber, and other natural scientists standing behind Germany's imperialism?

Einstein's words shed somewhat different light on the appeal "To the Cultural World." How many German scholars had unknowingly affiliated themselves with the president of the Goethe Society? And which ones? Could one of them be the conservative Nobel laureate in physics, Wilhelm Roentgen, who, participating in a survey in 1920, protested that he had signed the appeal "stupidly as some others have on the advice and sharp insistence of the Berliners," without having read it?[217]

Planck too told one of his fellow signatories after the war that he signed the appeal "without knowing its wordings, as at the time, September 1914, I was on a trip, and my children back at home presented the issue as so urgent that they signed on my behalf." When he later read the text, he was shocked and embarrassed. This, however, did not stop him from signing a declaration of similar content, the "Declaration of University Teachers of the German Reich" in the fall of 1914.[218]

Just two out of the ninety-three signatories immediately withdrew their signatures after they read the text. And merely a handful of other intellectuals followed their example during the war, although the editor-in-chief of the *Berliner Tageblatt* made a notable attempt in 1915 to provide a forum to explain the text in public to those who signed without knowing it.[219] Einstein had already conceded to his Dutch colleague Hendrik Antoon Lorentz that he did not believe the cosigners could be induced to withdraw.[220]

Planck himself informed Lorentz about his motive: Germany fought to preserve its highest and most sacred asset, namely its existence. This conviction alone caused him to support the statement of the German scholars. Indeed, he did not agree with every clause. Nevertheless, he thought that he had to join the statement, since it "could and should have just one purpose, to once and for all say in public, as seemed necessary, that German scholars did not want to separate their cause from the German government and German armed forces." After having done this, the scholars were left with a no more pressing task than to fully counteract the growing hatred among the nations.[221]

By this point, one of Planck's sons had already been injured and taken prisoner of war. Einstein took great interest in the fate of his colleagues. At the end of 1914 he wrote to medical examiner Heinrich Zangger that he almost traveled to Switzerland "in order to take the necessary steps to attend to one of Planck's sons who was injured and taken prisoner of war in France." But he had learned that the boy was out of danger and on the road to recovery.[222]

Why did Einstein stay in Berlin? Why didn't he return to Zurich? In the following years the Swiss city became a popular haven for pacifists and avant-garde artists, such as members of the Dada movement. During the war years, a number of Einstein's future companions, like the writer René Schickele, migrated to Switzerland, taking their magazines with them.

The internal and external conflict experienced by many German pacifists made residence in their homeland impossible. Einstein, however, had disengaged from the land of his fathers at a young age as he searched for an identity. Although basically stateless, he felt more Swiss than German and as a result of his experience and way of life, particularly immune to nationalism and the general decline in values. He remained a good Swiss, though he made a distinction between political opinions and personal connection.[223]

Since his separation from Mileva, Einstein no longer seriously considered returning to Zurich. In addition to Elsa, it was above all colleagues like Planck who kept him in Berlin, although he felt "extremely uncomfortable" there. He sensed that his presence alone had the effect of a silent reproach for some German scientists.[224] Einstein did not take part in the political debates of research veterans at the Academy, who regarded him as a dreamer at best on account of his pacifist position. As Einstein told the physicist Philipp Frank, his successor at Prague University who came to visit him in Berlin, "You cannot imagine how good it is to hear a voice from the world outside and speak about the world without inhibition."[225]

Frank was not allowed to bring any documents when entering the German Reich. "Every empty piece of paper was confiscated at the border, because there could be news written on it with invisible ink."[226] Moreover, for fear of espionage and an undermining of the will to fight, all international mail was examined by censors. Einstein could no longer write freely to his old friends Zangger, Besso, and

Ehrenfest. He had to mail his letters unsealed.

The post inspection agencies surrounded the Reich like a cordon sanitaire.[227] According to the rules, unreadable, overly long, or "wailing letters" should not be sent, and suspicious letters should be checked for secret writing. The censors were very thorough. "In a monthly maximum number of nine million pieces of mail, 1,700 cases of secret writing were discovered."[228]

Hoping the nightmare would soon be over, in the first months of war Einstein generally shut himself in. His withdrawal into the private realm seemed to turn out well at first. As long as he was left alone, he worked as usual, "without allowing himself to become infected by the mass psychosis."[229] Elsa, meanwhile, dedicated herself to volunteering at a soup kitchen. According to Einstein she cooked lunch daily "for a crowd of poor women."[230] He himself enjoyed her good cooking. Unfortunately, her domestic qualities were often the only thing he mentioned in his letters.

Einstein initially considered his marriage plans a closed matter. He was too anxious to be alone. He lived in his apartment in Dahlem like a bachelor. After sending most of the furniture to Zurich, he was left only with a desk, a blue sofa, and a couple of family heirlooms. His home now seemed empty. When the thought of his children sometimes came to him upon waking in the morning, it was as though he'd been stabbed by a dagger.[231]

In the fall Einstein started looking for a smaller apartment, closer to Elsa's home in Schöneberg, especially since the research institute that had been promised to him was not being built for the time being. The Koppel Fund had provided the means for building a Kaiser Wilhelm Institute for Physics in Dahlem, but with the beginning of the war the funds dried up and the building project fell to the wayside.[232] The Kaiser Wilhelm Society was also now assigning a good portion of its foundation capital to war bonds.

After a few excursions into quantum physics, Einstein was now more focused on his theory of relativity. He had become acquainted with many special cases in physics and appropriated complex mathematical tools in order to overcome the obstacles he encountered. But all the concrete examples and calculation methods had so far pointed only roughly in the direction of a new theory of gravitational physics. Some even indicated that his principle of relativity did not at all apply to his postulated hypothesis of the general theory of relativity.

Einstein gave two lectures at the Academy on his favorite theme, in which he presented his progress to date on the provisional conclusion of the theory of gravity. The response was modest. But as time passed, he increasingly believed he had created a conclusive, consistent theory. It would take a few more months before he had the inspiration for a decisive modification.

Political Declarations

Whenever Einstein went to the Academy or the university on Unter de Linden, he saw troops marching. For years he watched soldiers decorated with flowers on their way to the front, only instead of happiness and enthusiasm, ever more anxiety was revealed on their faces, and the image was etched in his memory.[233] In the weekly sessions of the German Physical Society, attendees rose to commemorate the fallen. And when he went to the Imperial Physical Technical Institute, where for a short time he discovered a love for experimentation, he was keenly aware of how much of an outsider he had become, due to the patriotic rapture of Berliners.

His correspondence always came back to the appeal "To the Civilized World." The "Manifesto of the Ninety-three" was the key to understanding Einstein's incipient engagement with pacifism and international understanding, which after the war would contribute substantially to his growing international popularity. The appeal signed by his closest colleagues propelled him to his first public protest. Einstein signed a counter manifesto and supported the "Manifesto to the Europeans," initiated by Georg Friedrich Nicolai.

Nicolai, a heart specialist and senior consultant at the Charité in

Berlin, whom even Kaiserin Auguste Viktoria called in for medical advice, was a notorious womanizer and good friend of Elsa. As one of the pioneers of electrocardiography, he was also the subject of conversation in scientific circles. Self-confident, he started, for example, an argument with a physics professor in order to teach him about the physical behavior and mathematical description of elastic membranes, for which he received support from no other than Max Planck.[234]

When Nicolai read the appeal "To the Civilized World" that Planck and others had signed, he was horrified. The elite German researchers spoke boldly, having been in a fighting mood that no nationalistic passion could justify. He immediately drew up a counter project, the "Manifesto to the Europeans," which he discussed with Einstein and the astronomer Wilhelm Förster. "Should Europa exhaust itself and perish slowly in fratricidal war?" It stated that never before had any war so completely broken the cultural community of life together as this present war. Technology had made the world smaller, the European states today seemed to jostle one another much as those ancient cities of each Mediterranean peninsula. It was therefore the duty of educated Europeans to at least try to prevent Europe from succumbing, for lack of international organization, to the same tragic fate that engulfed ancient Greece.

This raging war would hardly yield victors—it would leave behind only vanquished. "Hence it appears not only noble, but bitterly needed, that educated men in all countries exert their influence there so that—whatever the still uncertain outcome of the war may be—the conditions of peace would not be the ground for future wars." Instead, Europe must be unified. The technological and intellectual conditions were ripe for such an opportunity. Those for whom Europe was not just a geographical term but a vital cause, should join forces. An assembled European League should speak out and make decisions.[235]

The writers of the "Manifesto to the Europeans" advocated a lasting European peace and anticipated the idea of a League of Nations.

Hoping for wide support, the declaration omitted questions such as who was to blame for the war. The forward-looking manifesto contained the same political maxims that Einstein would come to stand for in the future: pacifism and international understanding.

The appeal circulated around the University of Berlin in the fall of 1914. But other than the three mentioned above and a colleague of Nicolai's, no one signed it. "Although the counter manifesto was discussed and accepted in a university lecture hall and a great number of copies were made available for the professors, including dispatches to private addresses that were met with friendly approval, yet no one was prepared to sign."[236] The project that marked the beginning of Einstein's political activity came to nothing.

Nicolai promptly announced that he was going to explain his opinions in a lecture at the university, "Germany's first intellectual bastion," where one could find those who were no longer idle.

The military authority banished Nicolai to Graudenz in the Prussian part of Poland to work as a guard at a military hospital. Thanks to his good connections, after a successful appeal he returned to Berlin and undertook a course of pacifist lectures in order to show the terrible consequences of war, whereupon he was once more called to military service in Danzig. Once there, he refused to take the oath of allegiance.

Nicolai's courage was exceptional. His fate, which will be discussed later, shows the consequences peace activists had to reckon with in those times. The German Peace Society, which had around ten thousand members who were active in hundreds of local groups, was completely silent during the first months of the war, largely for fear of repression. Well-known members had abandoned pacifism.

Einstein was not an activist as Nicolai was. He was Swiss, new in Berlin, and politically inexperienced, a loner who felt committed to science above all else. For the time being he made no public appearances. To overcome a feeling of helplessness, he was looking for contact with like-minded people. His pacifist position found expression first in his private correspondence. Most notably, his exchange of letters with Heinrich Zangger, whose entire work was published in 2012 by the American historian Robert Schulmann, longtime co-editor of *The Collected Papers of Albert Einstein*, reveals how the war shaped Einstein's political awareness and what worried him most.

Einstein felt that the particular calamity of the time was the technological equipment of the belligerent parties. The gap between scientific-technological progress and ethical-moral development was ever increasing. At the end of 1914, Einstein wrote to Zangger that the actual education of the masses was moving forward very slowly in relation to the rapid development of technology, so that "from now on the worst sort of misunderstanding is prevailing."[237] The deployment of new military technology would lead to a true annihilation. "I am therefore convinced that we must strive for a political organization on a larger scale that will act against an individual state the way the state acts against an individual criminal."[238]

Einstein, who always returned to fundamental concepts in physics in order to view them in a new light, here offers a surprising analysis that is as simple as it is shrewd: In the history of western civilization, violence has been transferred from the private sphere to the hands of the state. Since the state should protect individuals from criminals, it possesses a monopoly on the legitimate use of force. The downside of this evolution is the armament of the state, which the individual hardly perceives anymore. Times of peace are times of armament, during which under the scientific-technological development, the state requires precisely that which is suitable for strengthening its monopoly on legal force against its residents and with regards to possible interstate conflict. With that comes an immense potential for destruction.

At the onset of World War I only a very few had any idea what had been piling up in the arsenals during the previous period of peace. Einstein, too, was unaware. But now, as the dreadful destructive power of modern weaponry became visible for all, he came to the conclusion that a further escalation of violence could possibly be brought under control by a new political order. In the first months of the war the idea of a European League or a League of Nations was crystalizing as his central political idea.

At the same time, he was witnessing how an entire research institute was being put into military service. The war did not halt even at his workplace. The Kaiser Wilhelm Institute for Physical Chemistry in Dahlem, a particularly pleasant, quiet location for a muser like Einstein, was transforming into a major research facility for weapons of mass destruction. The head of this undertaking was Einstein's close friend Fritz Haber. Even the scientists from next door, whom Einstein saw in his first Berlin summer roaming the cornflower meadow in the Dahlem area, were no longer studying just the composition of leaves and flowers. They were designing gas masks.

6. The Inception of a Weapon of Terror

Suicide on the Home Front

At dawn on Sunday, May 2, 1915, Hermann Haber was startled by a loud bang. The boy dashed through the house, ran outside, and found his mother motionless in the garden, a pistol lying next to her that his father had been carrying around since he began wearing a uniform. The boy's mother, covered with blood, was still alive. "Her heart beat for another twenty agonizing minutes."[239]

Clara Haber's suicide shocked the twelve-year-old and his father, who was still dazed from the previous evening when the house was filled with guests he had invited to an impromptu celebration. Fritz Haber had just returned home from his first use of poison gas in Belgium, and as often happened at the time, was just passing through Berlin on his way to the Eastern Front. Herman had already gotten used to his constant absence.

For most of the schoolboys his age, the battles in the West and East were an exciting game. Journalist Sebastian Haffner visited a secondary school at Alexander Platz and later wrote about how the boys would play "trenches," study daily war bulletins, stick miniature flags in topographical maps of the war zones, offset German against Russian and French prisoners of war, and count downed airplanes and sunken submarines. "The military events at the time made life exciting and brightened up the day. A big attack with a five-figure number of prisoners and fallen strongholds and 'immeasurable spoils of war material' was celebrated as a holiday."[240]

On Sunday, May 2, 1915, a big offensive started in Galicia as German infantrymen, preceded by a constant barrage, stormed the Russian positions. The next day, bells tolled all over Berlin and public buildings were decorated with flags. Word came that the Germans took 160,000 Russian prisoners and seized 24,000 horses and 300 armored vehicles. The numbers published later were somewhat lower; the public immediately reacted with disappointment at "only" 21,000 Russian prisoners.[241]

This breakthrough of the Gorlice-Tarnów offensive was followed by the advance of German troops over several months. They drove out the Russian army from Russian Poland and in Kaunas alone seized more than thirteen thousand cannons and one million shells. These new numbers accelerated the fantasies of the Berliner newspaper readers and schoolboys. "The soul of the crowds and the childish soul react in a similar fashion."[242]

Herman Haber called in vain for help. His mother continued to bleed. His father too, with whose army revolver she shot herself in the breast, could do nothing to save her. She had put an end to her life with his service pistol.

The family tragedy got under everybody's skin—relatives, friends, and colleagues at the Institute. Just recently, Einstein had been giving "little Haber" math lessons. From January to Easter 1915, in the midst of his research for the theory of relativity, Einstein gave lessons in calculation and geometry to the boy, who due to illness could not attend school for a while.[243] Who would take care of the twelve-year-old now? Probably an aunt, for after his wife's suicide, the father of chemical warfare returned immediately to the front.

Fritz Haber wanted to be an officer through and through. The forty-six-year-old wore his uniform with an insane sense of responsibility. That very same evening he turned his back on the death scene in order to prove the effectiveness of his new chemical weapon in Galicia, where his regiment had been relocated. While young Hermann, nicknamed "H. Two" in the family circle, stayed in Berlin, "H. One" was heading off to assist with the planned large-scale attack in Gorlice-Tarnów, though he reached the front too late.[244]

On the evening of May 2, Fritz Haber called civil servant Friedrich Schmidt-Ott in tears: He must leave in half an hour for the headquarters. His wife could no longer bear living.[245] He used the same wording in a letter to his longtime colleague, the chemist Carl Engler. In this military letter Haber grants a rare insight into his soul and reveals his immense inner conflict, the rift between thinking and feeling.

For a whole month after his wife took her own life, he strongly doubted that he would be able to keep going, he told his friend. Now he must live through the endless frictions of the war with strangers; he had no time to look to his left or his right, reflect and sink into his feelings, driven by the horrible responsibility "that a missed day is paid for with blood.... This is the whip that is hanging constantly over my head." Only in between, sitting at the general headquarters chained to the telephone, he listens to his heart "and hears the words that his poor wife spoke then and there, and with exhaustion between orders and telegrams has a vision of her head, and agonizes."[246]

Einstein and Haber

Fritz Haber was born three years before the unification of Germany. "His childhood was spent in a phase of glowing enthusiasm for German arms and German unification," which he demonstrated even as a schoolboy, according to the historian Fritz Stern.[247] Yet he was barred from having a military career in his youth because of his Jewish origin. He suffered because of his Jewish ancestry and, like his wife Clara and his patron Leopold Koppel, converted to Christianity and also advised the new Berliner, Einstein, to convert to Protestantism. "Do that so that you will fully belong here."[248]

Einstein saw Haber's advice as symptomatic of the stance taken

Image 6: *Institute director Fritz Haber and the new Berliner Albert Einstein in July 1914.*

by many German Jews, that conversion was an admission ticket to the middle class. They aspired to fully integrate in German society, although that society mostly did not recognize Jews as equal. The industrialist Walther Rathenau, himself torn between German and Jewish identity, described this feeling very well: "There is a painful moment in the life of every German Jew that he remembers for the rest of his life, when he becomes for the first time fully aware that he came into this world as a second-class citizen and that no talent and no distinction will help him break free from this situation." However, a conversion and gaining social status and business benefits from denying his father's faith was out of the question for Rathenau.[249]

Einstein, who would later have many discussions with Rathenau about the position of Jews in Germany, vividly recalled scenes from his school days in Munich that made him "feel alienated." This discrimination found Einstein just as unprepared as Rathenau and many others, as he too grew up in an assimilated German-Jewish family. His parents neither went to the synagogue nor kept a kosher kitchen. Nevertheless, as the only Jew among seventy schoolboys he was exposed to some hostility, went through a brief phase of religious sentiments, and from then on was sensitive about any indication of antisemitism. For instance, he attributed the rejection of the letters he sent to German professors after he graduated from the polytechnic to antisemitism.

To Einstein, having Jewish roots meant being part of a community of fate. Unlike Haber and Rathenau, he neither finished his high school in Germany, nor was he instilled with imperial values during university years or military service. Looking back at earlier years, he would later explain that for a long time he was not aware of his Judaism, neither in Switzerland nor during a short interlude in Prague, where he represented himself as "nondenominational." Nothing in his life enlivened his Jewish sentiments in any way.

"This changed once I moved my residence to Berlin."[250] It was first in Berlin that Einstein saw the distress of many young Jews, above all those who, since the end of the nineteenth century, had been fleeing pogroms in Russia and had settled in the preferred Scheunenviertel (Barn Quarter). His harsh reaction when he was invited to St. Petersburg in May 1941 reveals how much their fate grieved him: "I am loath to travel, without a good cause, to a land where my fellow tribesmen are being brutally persecuted."[251]

Jews lived in the Scheunenviertel in Berlin as in a Polish shtetl, wearing a yarmulke and earlocks and speaking Yiddish among themselves. Assimilated Berlin families and Jewish academics set themselves decisively apart from these East European Jews, which Einstein ascribed to a deep sense of insecurity "that can amount to moral instability." The latent, often also open antisemitism subverted their self-esteem. At times the result was self-hatred. He saw "worthless caricatures of worthy Jews," and his heart bled at the sight.[252] Einstein thought it was overwhelming surrounding circumstances that shaped the "pitiable" baptized Jewish *Geheimrat* (Privy Councilor).[253]

Einstein saw in Haber one of those tragic figures, a converted Jewish Privy Councilor who wanted to blend into the community and at the same time stand out. His exaggerated need for recognition struck Einstein even at their first meeting. The way Haber received Einstein in Berlin reveals much about their relationship. The director of the institute used his good connections with the authorities and the industry to pave the way for the young genius whom he wanted in his proximity; he helped him search for an apartment and arranged his office and introduced him to the financier Leopold Koppel.

Significantly, their closest moments ensued when Einstein most needed Haber at his side: during his separation from Mileva and the children. Haber accompanied him as a loyal friend, made immediate contact with Einstein's cousin Elsa, and stood by her side too with help and advice. Einstein was profoundly grateful to him. Nevertheless, a few days after his painful separation from his sons he told Elsa that he associated with Haber for "exclusively objective matters."[254] This was the first break.

The turning point in their relationship came on August 3, 1914, the day Germany declared war on France. From that point, their lives took completely different paths. In the first months of the war Haber would become a key figure as go-between for research, the industry, and the ministry of war. He saw it as a matter of urgency to put scientific work into the service of the military and integrate research findings into military plans right from the start. With ever new agents of chemical warfare he contributed to the radicalization of the war that nipped in the bud all the efforts to reach a peace agreement.

Bread from Air

Haber's biggest scientific discovery involved a chemical element that is most common in the air we breathe and is essential for all plant and animal life: nitrogen. Air is composed of around 78 percent molecules of nitrogen. The two nitrogen atoms in this molecule are bound together extremely tightly. Only electrical discharge produced by thunderstorms and special types of bacteria, free living in the earth or in symbiosis with plant roots, are able to split the nitrogen molecule. The details of how the bacteria succeeds are to this day the subject of research.

Breaking, by technical means, the strong atom bonding in the nitrogen molecule and producing nitrogen fertilizer, was, in light of the growing world population, one of the most urgent problems on the threshold of the twentieth century. Haber and many other researchers tried to change nitrogen into nitrogen oxide using an electric arc, for which reason the Baden Aniline and Soda Factory (BASF) in Ludwigshafen, primarily a dye producing company, teamed up with him. Later, collaborating with the BASF again, Haber discovered a significantly more elegant and energy efficient procedure that enabled the production of nitrogen on an industrial scale. Thereby he became the man who created "bread from air." Not just bread, to be sure, but also explosives.

The first time he turned to nitrogen, Haber had in mind neither fertilizers nor explosives. He simply reacted to an inquiry from the industry. The Austrian Chemical Works wanted to know to what extent it would be profitable to synthesize ammonia from the two common chemical elements nitrogen and hydrogen. Haber knew that it was almost impossible to produce any reaction from nitrogen molecules. The expected actual yield of ammonia product was extremely small. Like all the leading chemists, he thought that ammonia production in this way and on such a massive scale was out of the question. Only because the company insisted, he set out to measure the smallest ammonia quantities at different temperatures.

His colleague Walther Nernst was ardently interested in the measurement results, as he wanted to illustrate his heat theorem with the same chemical reaction. Haber's results however did not coincide with his own calculations. Consequently, Nernst began to conduct experiments as well. He discovered that the formation of ammonia clearly increased by raising the pressure, but he could not verify

Haber's values, whereupon he notified his colleague.

Nernst was a luminary in this field. His word carried a lot of weight with researchers. Therefore, Haber returned immediately to his test tubes in order to review his measurements. At the next meeting of the German Bunsen Society for Physical Chemistry in Hamburg, Nernst boasted of his achievement and provoked a confrontation with the aspiring chemist from Karlsruhe: he politely requested that Haber use a method that would "actually give exact values for the bigger actual yield."[255]

Haber would never forgive his colleague for demeaning him in public. A quarrel flared up, dominated by mutual vanity. Afraid for his reputation, Haber, who could no longer rest easy, worked doggedly on the nitrogen problem, availing himself of Nernst's groundbreaking insight, putting the chemical reaction under increased pressure.[256]

Over the next few years his nitrogen experiments profited from his connections to the industry. In one case, through writing a report, he obtained an urgently needed high-pressure apparatus. Later, the German Gaslight Company AG, for whom he had already worked as a consultant, provided him with some particularly rare chemical substances such as uranium, tungstic acid, and osmium. It was the costly noble metal osmium, which the Berlin company had previously used for manufacturing lamps, that finally brought him the hoped for breakthrough.[257]

By this point, Haber had already tested many metals to see which surface atoms would easily combine with nitrogen. If a nitrogen molecule approached such a docking site, the strong chemical cohesion of the two nitrogen atoms would loosen. Using the appropriate metal catalyst, the nitrogen reaction could produce a more energy efficient process.

Until then osmium emerged to be by far the best catalyst for combining nitrogen with hydrogen to produce ammonia. In the summer of 1909, Haber invited an agent of the BASF to his laboratory to show him how ammonia dripped regularly out of his high-pressure device:

80 grams per hour. A scientific triumph!

From then on, his collaboration with the company continued under strictest secrecy; his results could only be published years later. The company was finally able to find a less expensive catalyst, better suited for industrial production than osmium, namely magnetite. To this purpose ambitious BASF researchers tested a whopping 2,500 different substances.[258] In Oppau, under the direction of Carl Bosch, BASF built the first ammonia factory that ran the now common Haber-Bosch process, and in 1913 it went into operation—one year before the beginning of the war.

Meanwhile, Haber had moved to Berlin, further climbing the career ladder. He held lectures, wrote professional articles, attended conferences, worked in the laboratory, applied for patents, and collaborated with the chemical industry.

His research activities led him through Germany, Europe, and to the United States. But all this came at a high price: ruined health, which constantly forced him to go on a weeks-long stay at health spas, and a marriage that had been shattered for years.

Fritz Haber and his wife came from Breslau. He was the son of a dye materials manufacturer; Clara Immerwahr, with whom he fell in love at dancing lessons, was the daughter of a chemist and, like him, wanted to pursue a doctorate, yet as a woman, she could only complete the secondary school exams by way of detours. Permission to attend university lectures in physics and chemistry depended on the good will of the professors.

Clara declined Fritz's offer of marriage, resumed her studies, and in 1901 took part in a German Bunsen Society conference as the first woman to be awarded a doctorate in chemistry.[259] A few months later she accepted the marriage proposal from the friend of her youth. After the birth of a son, she found to her great sorrow no way back into research. Occasionally she could help her husband with his book publications or lectures at schools for adults on "Chemistry in the kitchen and the house." Otherwise she reluctantly fulfilled the role of a mother and a professor's wife.

As Fritz Haber got lucky with the ammonia synthesis, Clara took stock of her pitiful marriage. "What Fritz has gained during these eight years—and more—I have lost, and what is left of me fills me with the deepest discontent." Even if she could partly blame her circumstances and her specific temperamental predisposition, still it was mainly Fritz's fault, next to whom "any character who does not force his way more recklessly at the other's expense will perish! And that is the case with me."[260]

The founding director of the Kaiser Wilhelm Institute for Physical Chemistry and Electrochemistry in Berlin was driven from one project to another by insatiable ambition. Haber found in the Reich's capital access to the highest circles. His colleague Richard Willstätter of the Kaiser Wilhelm Chemistry Institute said that in the hope of an audience with the Kaiser, Haber practiced subserviently walking backwards in his villa, breaking a Copenhagen vase in the process.[261]

Haber was never one of those scientists who can content themselves with having made one discovery. He was greatly engaged in the ensuing consequences. The war offered him new possibilities to implement his scientific knowledge and organize funds for research, for it soon transpired that without deploying new chemical processes for nitrogen production, the German army would run out of explosives as early as 1915. Nearly all the natural deposits of nitrogen were to be found in South America in Chilean saltpeter, namely sodium nitrate. Once England entered the war, all the Central Powers were cut off from its imports.

At the beginning of the war, research activities at Haber's institute almost ceased; most of his staff members had to enlist. With the help of his wife, a few rooms were temporarily converted into day nurseries.[262] During the first months of the war, Haber himself did everything within his power to meet the agricultural demand for fertilizers and the immense ammunition supply of the army. Synthetically manufactured ammonia could, as a result of chemical reaction with sulfuric acid, be converted into nitrogen fertilizer. But it was also possible to process ammonia through oxidation into artificial saltpeter for production of explosives.

Since the two processes were competing with each other, in the fall of 1914 Haber served as scientific advisor to both the agriculture and the war ministries and as the representative of the Baden Aniline and Soda Factory (BASF). The company had just opened its plant in Oppau and initially questioned whether to build an additional facility and invest in the Haber-Bosch process of ammonia synthesis, since it was not clear how long the war would continue and what market would open for nitrogen after the war. Haber kept the difficult negotiations going in the decisive phases until the basic conditions were so advantageous for the industry that the BASF could no longer eschew the risk.

Finally, the company gave its "nitrate promise" and committed itself to massively expanding the production of ammonia. Because of the constant rise in ammunition supply, specially designated factories had to give priority to further processing saltpeter. Haber's dramatic account reports that the agreement was reached just in time. "It succeeded at the last possible minute, as the saltpeter in the country has become so scarce that all the available saltpeter, in stock and transport containers and in processing, amounted to three weeks of ammunition supply."[263]

The agreement meant a big business deal for the BASF. "From January 1, 1915, until November 11, 1918, the company sold nitrogen compounds worth 414 million marks, of which fertilizers accounted for only a small part, namely 24.5 million marks," sums up Margit Szöllösi-Janze in her extremely detailed book on Fritz Haber and his connections to industry. After the war, the company easily adjusted its big new plants, funded by the state, to mass production of nitrogen fertilizer.[264]

Such numbers clarify that during the war the Haber-Bosch process served mainly for the purpose of producing explosives. Regardless, shortly after the war Haber was awarded the 1918 Nobel Prize in chemistry—but not because the Royal Swedish Academy of Sciences wished to honor solely the methodological progress achieved by

Haber. In the eulogy he delivered for Haber, Åke Gerhard Ekstrand, as president of the Academy, spoke not just about foundational research and pure theory; he wished to make clear that science determines the fate of nations. The world war had shown that each state must be able to produce essential chemical substances on its own territory.

To elucidate this point, Ekstrand finally referred to the production of fertilizers in the "German Haber factories" and spoke about a triumph "in the service of mankind."[265] Not a word about explosives. Haber too, followed this path in his ensuing lecture.

The split that occurred here is significantly characteristic. Even now, scientists apply for projects and funds, priding themselves with the positive results their discoveries have for society. Often, the consequences and possible damage for humans and the environment are not even mentioned. The latter is notable in Haber's case.

The Haber-Bosch process accounted for around 50 percent of the entire German nitrogen production in the last years of the war.[266] Without this stock the Central Powers (Austria-Hungary, Germany, Bulgaria, and the Ottoman Empire) would have had much less readily available ammunition and might have been compelled to an early peace agreement.

The Raving Barbarian

The announcement that the Nobel Prize was being awarded to Haber immediately after the war triggered waves of international protest. The rage was mainly about the unscrupulousness with which Haber promoted the development of new weapons of mass destruction. Can someone who has the death of countless people on his conscience be presented to the world as a benefactor of humanity and be honored with the highest award that science can give?

Einstein did not express his opinion on this in the preserved letters and documents. While Zangger complained that the Nobel Prize was not awarded to him but to the "poison fanatic" Haber, Einstein cloaked himself in silence.[267] Why? What did he think about the researcher who stood by him when he moved to Berlin then evolved into the mastermind of poisonous gas? What are we to make of the fact that their relationship spanned over two decades, from the German Empire through World War I, and from the Weimar Republic up to the National Socialist rise to power?

Haber's admiration for his colleague during this eventful time is emphasized in a letter he sent to Einstein for his fiftieth birthday, as the latter was at the peak of his fame. Haber wrote, "Of all the great things I have experienced in this world, the content of your life and work are the most profound. A few hundred years from now, the simple man will identify our time as the period of the world wars, but the educated will link the first quarter of the century with your name.... Then all that will remain of the rest of us is the connection between us and the great events of the time, and, I think, it will not remain unmentioned in your sufficiently detailed biography that you had me as your partner for more or less pointed remarks about the Academy, and more or less bad coffee."[268]

One looks in vain in Einstein's correspondences for similarly unbridled appreciation for Haber. Again and again one comes across critical undertones, comments about his vanity, his conformity, and his patriotism. Even when Einstein had immigrated to the USA and shortly afterward Haber likewise became a refugee, Einstein could not hold back and told him how very happy he was that "Your earlier love for the blond beast has cooled off a bit."[269]

Even the condolence letter he sent to Haber's son Hermann in 1934 is ambivalent. It says that Haber had been his most ingenious, versatile, and helpful friend. But toward the end of his life Haber had to experience the bitterness of being abandoned by the people of his circle, "a circle of which dubious qualities he was aware, yet which meant a great deal to him." He saw in Haber's fate "the tragedy of German Jews, the tragedy of unrequited love."[270]

One of Einstein's rare statements in writing about Haber to a third party is found in an exchange of letters with the physicist Max Born. Einstein had a close relationship with Born; he offered to address him with the personal pronoun "you" (Du) instead of the third person (Sie)—something he never did with Haber. Born, a "great fellow," was one of the few German scientists to show some backbone; in the summer of 1915 he refused to collaborate with Haber on his research of poisonous gas, because in his opinion it was necessary to put limits on the permissible—otherwise everything would soon be acceptable. Immediately after the war Einstein wrote to him that Haber, who in his misery threw himself on him, wanted to wrest truth from nature with violent methods. "He is sort of a raving barbarian, but then an interesting one at that."[271]

Born considered this an accurate assessment, which he illustrated with a memorable experience. He once had a lively discussion with the "raving barbarian" in his room, constantly interrupted by assistants, doctoral students, and engineers, who all wanted something from the head of the institute. "Finally, the door opened without a knock, whereupon Haber, raging, grabbed a glass inkwell and flung it in the direction of the door, where it shattered, staining the wall and the door with ink. It was his wife, though, standing at the door. She disappeared in horror, and we continued with our work, as if nothing happened."[272]

Continuing with his work, as if nothing happened, was a characteristic trait of Haber. Nothing blew him off his course: not his frail health, which he often strained beyond human measure; not his wife's desperation nor any explosion, such as the one that had occurred on December 17, 1914, in Berlin. That detonation, which took place four and a half months before Haber's wife committed suicide, was the first deadly accident related to war research in the Kaiser Wilhelm Institute for Physical Chemistry and Electrochemistry—an accident that almost cost Haber his own life.

At this point in time the chemist was once more in competition with Walther Nernst, who had been on duty at the front as a driver. In September 1914, Nernst, the "gasoline lieutenant," experienced the unsuccessful Battle of the Marne and the retreat of the troops. After this turning point, the armed forces of the Central Powers and the Allies held each other in check. Trenches, barbed wire fences, and machine guns, which fired several hundred bullets per minute, averted almost any attempt to gain ground.[273]

Even before the mobile warfare became static, Nernst had tested different chemicals that irritate the eyes and the respiratory tract. The Chief of the General Staff, Erich von Falkenhayn, had instructed him to find a substance for fighting resistance in houses, farms, villages, and other entrenchments, whereupon Nernst immediately contacted the industrialist Carl Duisberg, a potential producer of the desired warfare agent.[274] Together, starting in the autumn of 1914 at a shooting range near Cologne, they examined the effects of various combinations of dianisidine.

Duisberg had entered the military business with great enthusiasm. However, when on October 27 around three thousand dianisidine-filled shells were fired in Neuve Chappelle, the substance evaporated so fast that the French soldiers hardly noticed it. The attempt to force the opponent to abandon his position failed.

Right from the onset of the war, French troops had been firing cartridges containing eye irritants that were obtained from police stock. Corresponding hand grenades had been developed in France as well, but they had hardly any effect in the open field. After reading a newspaper report about the use of chemical weapons, Einstein remarked wryly: "This means they stank first, but we can stink even better."[275]

Following the failure, Nernst and Duisberg were on the lookout for an alternative and tested various irritant gases on rabbits and on themselves. Meanwhile, they had competition from several sides.[276] Among others, Fritz Haber was in search of less volatile and therefore more effective "stinks," as they were called in military circles. Up to that point, at the behest of the army he had developed a frost protection agent for gasoline and researched explosives and substitute materials. Now he expanded the laboratory research at the institute into chemical irritants and warfare agents.

On December 17, 1914, he and his associates at the Kaiser Wilhelm Institute were experimenting with cacodyl chloride. This highly poisonous substance had been used over the previous days in firing trials conducted in Kummersdorf, south of Berlin. On the seventeenth, Haber left the laboratory for a moment, and two professors, Gerhard Just and Otto Sackur, resumed the test. No sooner had they added a little methyldichloramine to the starting substance than the aggressive blend blew up. A terrible explosion shook the building. Just lost his right hand, and Sackur was fatally injured.[277]

Haber escaped the disaster by coincidence, because he had left the room. He stood petrified, as his wife Clara, who heard the explosion, hurried to the scene. Years later the institute's engineer Hermann Lütge remembered that Clara was the only one in this situation who tried to help Sackur, who was completely mutilated and in deadly agony, before the ambulance arrived.[278]

The next day, Haber opened one of the German Physical Society meetings over which he presided and which Einstein also attended, with an obituary for Sackur. The physicist Lise Meitner, from the neighboring institute in Dahlem, recalled how Haber fought to hold back tears during his brief address.[279] Sackur had become a department head in Haber's institute just nine months earlier. The institute director advocated that the accident victim be put on equal footing with the 240,000 war victims that the German army was already lamenting at this time.[280] Hence Sackur's wife and daughter could at least have certain social security benefits.

Einstein held Sackur in esteem for his work on quantum theory; a year later he dedicated a lecture to his peer, who was also his contemporary.[281] We do not know if Einstein experienced the accident at close range. We do know that his office at the Kaiser Wilhelm Institute was not impacted. Since giving up the apartment in Berlin-Dahlem, which was too big for his needs, and moving what little furniture he had left to Wilmersdorf, he went to the institute only sporadically. His new home on Wittelsbacher Strasse 13 was about a fifteen-minute walk from Elsa's house and just a few stations away from Haber's institute by street car or underground. His correspondence suggests, though, that from this point forward he stayed predominantly at home, where nothing distracted him from his scientific work, incubating his physics.

The First Year of War: A Gloomy Balance

Berlin's leading natural scientists still liked Einstein's company. He accepted Nernst's invitation for Christmas, as families mourned their relatives who had fallen in the first year of the war. The chemist, who had already received the Iron Cross 2nd Class military decoration for his service in the war, lived with his family in a magnificent house from the "founders' period" (the late nineteenth-century entrepreneurial boom) near the Potsdamer Platz. As Nernst's daughter Edith later recalled, the mood was gloomy. She had lost her brother Rudolf, and her father had nothing good to report about the course of the war on the Western Front.[282]

Thanks to his activity as dispatch driver and his contact with high-ranking officers, Nernst had an overview of the military situation. Chief of the General Staff Erich von Falkenhayn had decided to go into positional warfare. He no longer believed in a successful offensive in the West; the losses among the German ranks were too heavy. The Germans were threatened by "attrition warfare."[283] Falkenhayn wanted to gain time with positional warfare in order to fully "harness the power of science and technology for the war."[284]

Shortly before Christmas, Nernst contacted the Berliner chemist Emil Fischer, who had provided him with pure hydrogen cyanide for military purposes. During a personal meeting with Falkenhayn on December 18, Fischer found out how dissatisfied the chief of the general staff was with the effect of the present "stinks." "He wants something that will render people permanently unfit for battle."[285] According to his own statement, Fischer knew of a substance with such a deadly and enduring effect; however, he was fearful the Germans might shoot themselves in the foot, because it was their enemies who had access to the necessary raw material.

Nernst likewise thought Falkenhayn's objective was unattainable. On Christmas, the sociable scientist asserted "to his family and friends' dismay" that the war could no longer be won on two fronts. "It was neither defeatism nor lack of patriotism" noted his student and biographer Kurt Mendelssohn, "but the cold critical analysis of a scientist."[286]

In the meantime, high-ranking military staff urged the chief of the general staff to limit the war on the Western Front to defensive war and attack on the Eastern Front, where the allied Austrian-Hungarian troops suffered a severe defeat. In their view, the last hope to escape a lasting double envelopment lay in pushing back the Russian forces with a spring offensive and forcing a separate peace treaty with the Russian Empire. For the time being, Falkenhayn did not want to significantly weaken the Western Front.

At Nernst's house Einstein likely found out much more about war-related events than he could learn from the censured newspapers that fueled expectations of victory. Immediately after the holiday he wrote in a letter to Heinrich Zangger that the world appeared to him as a "madhouse." "What drives people to kill and mutilate each other so furiously?" he wondered. Einstein suspected a strong masculine drive, while "someone entirely dispassionate as I is perceived by others as damaged."[287]

Although his views made his colleagues think he was "damaged," Einstein did not hide them. He held on to moral values that had been declared invalid by his environment. Disgusted by the war's abominations, he preferred to withdraw into his brooding. He told his son Hans Albert in 1915 that he spent entire days in his small apartment and sometimes even cooked himself lunch.[288]

He hadn't seen his children for half a year now. He missed his "Albertli" and "Tete." But just as before, he deemed it right that the boys not grow up in a house where their parents confronted each other as enemies.[289] He was especially happy that his eldest was making progress in mathematics. "That was my favorite pastime when I was just a bit older than you, around twelve years old. It would have brought me great pleasure to be able to teach it to you." Unfortunately, that was impossible. Instead, he now gave private lessons to the sickly Hermann Haber.[290]

Deadly Chemistry

Contact with Dahlem did not break off immediately after Albert's move, but it did change because, among other things, Hermann's father was no longer an available "partner for more or less bad coffee." Haber was now seen less and less in the scientific circles that Einstein felt he belonged to. Fritz Haber no longer had time to deliver lectures to the German Physical Society. And although he became a member of the Prussian Academy in December 1914, it took two years before he participated in a meeting.[291] He also disappeared from Einstein's personal milieu. His name, previously mentioned often, suddenly no longer surfaced in Einstein's extensive correspondence.

Research in the history of science has not taken note of how little evidence there is of Einstein and Haber's relationship during the war. Such evidence turns up again in relation to the founding of the Kaiser Wilhelm Institute for Physics in October 1917, with Einstein as its director and Haber on the board of directors, along with Planck and Nernst. After the war, Haber finally found out that Einstein was toying with the idea of leaving Berlin. His reaction is revealing.

"Our life has been colorful, since we became closer at the natural scientists' meeting in Karlsruhe," he began his letter from July 20, 1919. "But I think that if the war years moved us apart, they nevertheless endow me with the moral right to ask you to inform me of your intentions, in order to negotiate a return to Zurich." Should the reason be of an economic nature, Haber continued, there would be a satisfactory solution. It would greatly contradict the wishes of colleagues and the state "that we in Berlin should lose you."[292]

The manner in which Haber opened his letter, as well as other sources, supports the conclusion that the two researchers had hardly anything to do with each other during the war. Contradictory to representations in some biographies of Einstein, during the war Haber was by no means Einstein's "best friend." Rather, the war that tore families, friends, and acquaintances apart drove a wedge between the two.

In 1919, Haber, along with Planck, made every effort to keep Einstein in Germany. As he had done before the war, he served his revered colleague by using his still excellent connections to the patron Leopold Koppel and to the science administration to help Einstein with a considerable salary increase and better equipment for his institute. All this after consulting Elsa Einstein, with whom Haber now met from time to time, in order to discuss financial issues. "You know that Haber is a good friend of mine," Elsa told her Albert while he was on a visit to Holland in the spring of 1920. "We understand each other well and have something in common: we both want only that which is good and nice for you."[293]

But back to the war years. The war separated the two researchers even in its first months, because Haber, as mediator in the matter of producing explosives, was always on the road. As Haber biographer Margit Szöllösi-Janze elucidates, there were no official bodies for this matter. Haber personally had to establish contacts to the industry. He traveled from one location to another. How far apart Haber and Einstein's paths became was to be determined only at the turn of the year. From then onward, each incrementally moved in circles the other had no access to.

At the beginning of 1915, Fritz Haber presented the chief of the general staff with the prospect of what he had hoped for: a means for making people "permanently unfit for battle." Less than three weeks after the explosion in his institute, Haber had planned a chemical attack in the Western Front—not with a combat agent that is added to the shot, but with a gas, heavier than air and released from its own pressurized gas cylinder.

Haber became convinced that chlorine was the substance of choice for such a purpose. Chlorine was available to the German industry in large quantities. At the liquid phase it could easily be filled into transportable steel bottles. Once it was brought to the front battle line, Haber wanted to open the pressurized gas cylinder at the strategically and meteorologically favorable moment and let out the poisonous gas so that it rolled toward the enemy trenches like a dense, solid cloud.

In a language typical of him, the chemist explained that in positional warfare, the defender is superior to the attacker. "First: humans offer the present machine gun and field artillery pieces a meeting area that, given the number, rate of fire, and penetrating power of this weapon, is unbearably big; second: an easily established shelter (trenches) gives a long-range cover against these weapons, because small, rapidly flying iron parts cannot break through sandbags and earthwork. At the same time, this situation creates the need for chemical weapons, for the enemy as well as for us."[294]

Back in January Haber had gone to Cologne-Wahn and a little later to Belgium in order to test his blowing process under realistic conditions in an open field. We find out from Max Bauer, the artillery expert of the Supreme Headquarters, that he and Haber nearly poisoned themselves: "The gas flew off according to regulations, then the devil plagued us and we both rode 'by way of trial' into the moving cloud. For a moment we lost our orientation in the chlorine fog, an insane cough began, the throat felt constricted…in the nick of time the cloud dissipated, and we were saved."[295] Haber, though, used this episode later in order to downplay the effect of the chlorine gas. After numerous field and animal experiments with dogs and cats it was clear to all those involved that chlorine, in the quantities planned by Haber, is a deadly poison from which their own troops must first be protected.

Many high ranking military commanders were not impressed by the intended warfare. It was considered unchivalrous and distasteful.

In retrospect, General Berthold von Deimling wrote that it stuck in his craw "to poison the enemy like rats." But all thoughtful considerations finally stilled in the face of the noble objective of a German victory. Moreover, the commander-in-chief of the German Sixth Army, Rupprecht von Bayern, feared the enemy would soon resort to similar means. "Thereupon I was told in reply that the chemical industry of our enemies was not capable of producing gas in the required quantities."[296]

Haber had considered such questions from the beginning. He persuasively brought to bear all the knowledge he had gained through experimentation. France was, as regards chlorine, entirely dependent on import; England had at its disposal only a fraction of the German chlorine reserve. This, however, could not protect the German Reich from having the two countries immediately ramping up chlorine production and half a year later retaliating with the same warfare agent and benefiting by the prevailing west wind. Chemists the likes of Emil Fischer, who foresaw this development, wished Haber "failure from the bottom of my patriotic heart."[297]

Despite such criticism Haber carried on purposefully with his plans. His assistants described him as a great organizer and a person who commanded respect. Max Born himself, who had declined Haber's chemical warfare and broke all personal contact with him during the war, spoke of him as a fascinating personality, "full of life and energy, with exemplary, though old-fashioned, manners, an alert fast-thinking mind, and a wide range of interests."[298] Haber's longtime pen friend Richard Willstätter, 1915 Noble laureate for chemistry, described Haber in writing as "Primarily cool-headed, intrepid, death defying." That very same year, 1915, Willstätter supported Haber with the development of gas masks, and then Leopold Koppel's company, the German Gaslight Company, went into their production. Haber's network functioned outstandingly.

Prior to Willstätter, Haber had already recruited several other future Nobel Prize winners to his gas unit: James Franck, Gustav Hertz, and Otto Hahn.[299] The chemist Hahn described in his memoirs how he was summoned by Privy Councilor Haber, who was staying in Belgium on behalf of the ministry of war. "He explained to me that the only way to break the standoff in the Western Front was with a new weapon—aggressive and poisonous gas, above all chlorine, which would have to be released at the opponent from the foremost position. When I objected, saying that such warfare is a violation of the Haag Convention, he reckoned that the French had broken the first ground—even if insufficiently—namely with filled rifle ammunition. In addition, countless human lives would be saved, if the war could be terminated faster in this manner."[300]

Falkenhayn did not have the means in 1915 to win the war in the west by breaking through the front. Instead, he counted on the psychological effect of the chemical gas, which would transform the environment into toxic surroundings: a new kind of terror weapon. The military commander expected from the scientist a chemical substance that would make the enemy's shelters uninhabitable for the longest time possible.

Instead of seriously examining the impact this large-scale use of gas would have on blurring the boundaries of warfare, Haber wanted to demonstrate the superiority of German research. The desire for recognition and power drove him at least as much as his patriotism. He mobilized the smartest minds for this chemical warfare and gradually expanded his institute in Dahlem. War, for him, was also a continuation of science by other means.

Otto Hahn was one of the many leading researchers who succumbed to Haber's arguments. In a letter from March 1915, the physicist Lise Meitner encouraged her colleague Hahn: "I believe I have an idea of what you are engaged with and can very well understand your thoughts. And yet, this time, you definitely have the right to be an 'opportunist.' First, you were not asked; second, if you do not do this, someone else will; and above all, every means that can shorten this war is blessed."[301] In the deceptive hope of a quick German victory, she too argued past the cruel reality of the chemical warfare that led to the next unmanageable arms race.

While the intensive preparation for using poisonous gas was beginning under the alias "disinfection," Clara Haber visited her husband at the firing range in Cologne-Wahn. She was not the first woman to visit the place. But unlike Emma Nernst, who stood by her husband's side there, Clara Haber opposed her husband's plans unequivocally.

"We stayed in the Dom Hotel in Cologne," recalled one of the gas pioneers. "Privy Councilor Haber was accompanied by his first wife, a nervous lady, who already back then was sharply opposed to the Privy Councilor's intention of accompanying the new gas troops to the front."[302] As a chemist she knew the dangers to which her husband was exposing himself. The animal experiments she had seen left no doubt as to the threat to those who were caught unprotected in a gas cloud.

Against her will, in February her husband traveled to Ypres, where together with his assistants he positioned thousands of gas cylinders and secured them against the threat of hand grenades. The planned deployment was delayed week after week due to weather conditions. An artillery hit destroyed quite a few chlorine cylinders, killing around twenty soldiers from within the German ranks. "It was only from that day," according to one of Haber's assistants, "that General Deimling became convinced of the terrible effect of gas weapons."[303]

The chemical warfare agent intended to cause trouble for the enemy strained first all of the German nerves. Also Clara Haber's nerves. Her husband, who was waiting at the foremost front for the east wind, took very little notice of what was troubling her after she returned to Berlin, knowing about the imminent gas warfare. Following the description by the chemist Paul Krassa, who was related to her and whose wife was in close contact with Clara at the time, she "was desperate about the atrocious consequences of the gas warfare, having seen with her own eyes the preparations and tests on animals."[304]

Last Chance for Diplomacy

Einstein, who since his arrival in Berlin had often been Haber's guest, lost sight of his colleague; however, he might have seen Clara a few times during these weeks. Historical sources do not mention anything in this regard, except that he called her son Hermann a "clever boy" and that the private lessons Einstein gave him were not entirely fruitless.

Unlike the unfulfilled chemist, whose life as a helpless witness at her husband's side had become unbearable, Einstein found refuge in his study of physics. Science remained his anchor and consolation. His friends abroad were happy about his unabated creative energy: Zangger told him that Leonardo da Vinci likewise developed his deepest thoughts in a time of war.[305]

Einstein himself was shaken by the consequences of the war. He oscillated between pessimism, which at times became gallows humor—"I am now beginning to feel at ease with the present insane rumpus, consciously detached from all the things that engage the crazy public. Why shouldn't one enjoy life as a staff member in a madhouse?"[306]—and a growing willingness for opposition. His name and Elsa's, entered as "Frau Einstein," appeared in the meeting minutes of a new pacifist organization, *Bund Neues Vaterland* (New Fatherland League), for the first time in March 1915, and henceforth regularly.

Perhaps it was the Telefunken Company director, Graf George von Arco, who introduced Einstein to the nonpartisan association, which included Ernst Reuter, later the mayor of Berlin. The director was the former cavalier officer, prominent equestrian and author Kurt von Tepper-Laski. Diplomats such as Fürst Lichnowsky, former ambassador in London, and Graf Unico von der Gröben, former embassy councilor in Paris, stood by the association's side with advice.

The organization wanted to influence German foreign policy and establish contact with partner organizations and leading figures abroad. All members were called upon to take an active part. The high membership fee of at least fifty marks a year was used for the distribution of brochures and memorandums.

The New Fatherland League was formed so that the debate over German war objectives would not be left to the well-organized political right. In August 1914 the Kaiser and the Reich Chancellor declared that Germany was waging a defensive war, in order to legitimate the declaration of war on France and Russia and the immediate invasion of Belgium. Subsequently, German troops, which had suffered heavy losses, were pushed forward into enemy territory and were soon at the end of their tether.

"Two of my brothers lost their young sons in the battlefield," Planck lamented in a March 28, 1915 letter to his Dutch colleague Hendrik Antoon Lorentz. "My own sons are still alive, but one is wounded and a prisoner of war; my daughter works at a military hospital. There is hardly a German family that has escaped grief. Where is the equivalent of such heartache?"[307]

The many fallen, wounded, and prisoners of war were reason enough to end the war in the fastest possible way. But in the eyes of many contemporaries, they were instead the reason to continue with it. Hence after its sad soul searching, Planck's letter takes a quasi-religious turn: "And yet, the more than one thousand years of German history has not seen the German people so united. Should it be a bad thing that it brings about such a willingness to make sacrifice, such pure, sacred enthusiasm? I cannot believe that."[308]

From a historical perspective, the first months of 1915 seemed perhaps the last chance for diplomacy to end the slaughter in Europe. The fronts were immobilized and the terrible consequences of continuing with the war foreseeable. "It is one of the most remarkable paradoxes of historical moments when so much physical courage was raised for combat operations that the moral courage for an initiative that would rise against the stream and try to hinder them was extremely minor if not altogether absent," notes the political scientist Herfried Münkler. It certainly took courage to openly admit in this situation that all the suffering had been in vain. Among those responsible in

Berlin, Paris, and London, there was no one who had been willing to do so. "For that reason too, the war continued, because the politicians feared altercation at home."[309]

After the censured press spread excessive hopes of victory, the German people expected not a negotiated but a victorious peace, which Germany would dictate to its enemy. German troops stood deep in "enemy territory." Meanwhile, alongside the *Alldeutsche Verband* (Pan-German Association), the leading trade and farmers' associations demanded blatantly that territories conquered with "German blood" not be handed back.

The industrialist Carl Duisberg's change of heart, after being moderate at the war's onset, is typical. On March 3, 1915, he wrote to the Reichstag deputy Gustav Stresemann: "As unpleasant as it may be to annex Belgium for political reasons, perhaps as Crown land or as a colony of the German Reich, and despite the many enemies the Reich will thereby incur, we are bound to accept and come to terms with this inconvenience for military and economic reasons, since in my opinion it would be a big mistake not to draw into the Reich's sphere of interests this region that has become economically and agriculturally so important through its wealth of coal, the resulting economic working industry, and its advantageous location."[310]

A week later, the industrial and farmers' associations submitted to the Reich's chancellor, Bethmann Hollweg, their claims, which they reinforced in May. They insisted on incorporating Belgium into the German Reich. Furthermore, the French channel coasts would have to be annexed. There was also talk of incorporating East Russian provinces, new colonies, and high reparations.[311] A similar petition was submitted to the chancellor again in the summer, this time signed by 352 German professors, around 250 artists and writers, more than 150 ecclesiastics, as well as senior officials and business leaders. Bethmann Hollweg did not feel strong enough to stop the annexationists.[312]

In this political constellation, the New Fatherland League, which Einstein joined, became an important opponent to the annexationist movement.[313] The League was primarily concerned with backing up the politicians who made efforts to reach a negotiated peace.[314] In the memorandums it distributed to officials, which were partially reprinted by the press, the League expounded on why annexations had to be renounced for the purpose of a lasting peace in Europe. Einstein was directly involved in the composition of at least one of these appeals.[315] His regular participation in the meetings and his collaboration with the League's committees during the following months were documented in the meetings' minutes.[316]

As a Swiss citizen Einstein was above all interested in improving German-French relations. In March 1915 he addressed Romain Rolland, the French writer and pacifist who had been living in exile in Switzerland. From his exile in Genf he supported the work of the International Red Cross, and with newspaper articles and open letters called upon European intellectuals to raise their voice against the greatest catastrophe in hundreds of years. Einstein wrote that he had been informed through the New Fatherland League how courageously Rolland put his own existence at risk by calling for an end to the disastrous, deplorable state of affairs between the French and Germans. For that he wanted to express his complete admiration and esteem. "I offer my weak power in case you should think I can serve you as a tool, either with my place of residence or my connections to the Germans and foreign representatives of exact sciences."[317]

The League aroused special attention in April 1915 when a few of its members—with the consent of the ministry of foreign affairs—participated in an international pacifist congress in Den Haag. On the brink of the conference, the Dutch chairman himself contacted the German delegation to offer to come to Berlin for an exploratory talk. But the peace initiative came to nothing, as German politics once again kept a low profile. The German press immediately torpedoed the attempt and the peace initiative came to nothing. The *Norddeutsche Allgemeine Zeitung* (North German General Newspaper) wrote that no judicious mind would think of betraying the military situation, favorable for Germany, in favor of a premature peace agreement.[318]

The military remained at the ready. On April 22, 1915, at precisely 18.00 hours, the German gas pioneers that gathered around Fritz Haber in Ypres, opened the valves of 1,600 large and 4,130 small steel cylinders. This time the wind blew in their favor—northeast. A six-kilometer-wide yellowish cloud of smoke formed, drifting toward the French position.

As the alarmed Brigadier General Jean Henri Mordacq rode to the front shortly afterward, he found his army already completely disbanded. "Fleeing everywhere, French territorial troops, Africans, shooters, Zouaves [North African French troops], and artillerymen without weapons—distraught, their clothes removed or half opened, tearing at their collars, half blind—ran as madmen into the unknown, crying out for water, spitting blood." Others rolled on the ground gasping in vain for air. The general said he had never witnessed such a sight of complete dissolution.[319] Since the French soldiers were unprotected when the chlorine gas cloud rolled toward them, just over two meters high in some parts and thus very dense, it killed and wounded many. Estimates fluctuate between one thousand and five thousand killed and around three times as many wounded.

"Isn't this a gruesome way of fighting?" wrote the historian Gustav Mayer two days later to his wife in Berlin from German-occupied Brussels. One had to admit that the enemy would just as unthinkingly resort to all possible means of destruction. "It is hideous, though, that science, which otherwise strives for the preservation and increase of human life, is now put in the service of destruction everywhere."[320]

With the first extensive deployment of chemical mass destruction, a new threshold of propensity toward violence was transgressed. In the team of advisors to the Reich Chancellor, one spoke of the "collapse of international law." The chlorine vapors were never again to be banned from warfare.[321] An official debate of international law, however, never took place. On the French and British side, the question "Can we also do that?" led to immediate measures for retaliation.

Carl Duisberg raved about the "chlorious" victory in Ypres.[322] A victory that admittedly tore a gap in the French front but ultimately brought no territorial gain. In retrospect, the chemist Otto Hahn said that "already at this time there were not enough reservists available who could secure and exploit the breach in the enemy lines."[323]

To his great satisfaction, Haber was promoted to captain and immediately obtained the nickname "Colonel Stinker."[324] That very same week he set forth with part of his gas troop to the Eastern Front, where he was supposed to support the offensive in Gorlice. The Russian forces were pushed back from a region that the German annexationists had targeted for a new settlement project.

On his stopover in Berlin on May 1, Haber celebrated the military victory in Ypres. The short stay ended on the morning after the party with family tragedy: the suicide of his wife Clara. Her suicide letter disappeared. Inquiries in the 1950s by her surviving family members and former associates brought to light contradictory information. It is hard to evaluate how reliable the statements are. But the timing of the suicide as well as the circumstances indicate that Clara Haber put her husband's gun to her head in part, if not primarily, as a protest against his leading role in the chemical warfare.

When Einstein later called Haber a tragic figure, we can presume that Clara's suicide and his concern for "little Haber" shaped this image. But in no way do Einstein's letters shed light on the suicide. To his wife, Mileva, who had lived with the Habers a few times, he wrote merely a single sparse line: "Frau Haber shot herself two weeks ago."[325]

Part III: The Gravitational Field

"I have once again perpetrated something in the theory of gravitation that puts me in danger of being interned in a madhouse."[326]

— Albert Einstein

Race Toward a Theory of Everything
"The most exciting time of my life"

"Einstein is still young, not very big, with a wide, long face, thick mane, somewhat frizzy and dry, deep black graying hair that stands over a high forehead, with a fleshy, prominent nose, small mouth, thick lips, cropped moustache, full cheeks, and a round chin."[327] That is how the French writer and pacifist Romain Rolland described the thirty-six-year-old researcher after their first meeting on September 16, 1915. They sat together the entire afternoon on the hotel terrace by Lake Geneva—Rolland, "Europe's conscience," and the yet largely unknown physicist. In the midst of swarming bees, drawn to the flowering ivy, the two talked about the role of European intellectuals in the ongoing year-old war.

The meeting with the physics professor left an unusually wide track in Rolland's diary. Einstein's "lonely, happy, absolute intellectual independence" impressed him. Einstein's opinion of Germany seemed uninhibited. "No German commands such freedom. Someone different would have suffered from feeling isolated in his thinking. He does not. He laughs."[328] For instance, he laughed openly at his fellow professors, who after every senate session at Berlin University, met at a pub and started their conversation with the question: "Why does the world hate us?"[329]

Einstein's laughter and acerbic remarks occasionally left the impression that he was an aloof observer, indifferent to everything. But sharp irony was one of Einstein's only means of escape. As Rolland noted, he could not but give his serious thoughts a humorous form. Einstein was Jewish, which according to Rolland explained his worldly opinions and the mocking character of his criticism.

This character sketch by the 1915 Nobel laureate in literature was the result of an exceptional situation for Einstein: his first visit abroad since the outbreak of the war. As soon as he crossed the border to Switzerland, where the various nations still lived together harmoniously, where just as before, newspapers from Germany and France,

Austria-Hungary and Italy, lay peacefully side by side in the kiosks, Einstein took a deep breath. Finally, he could speak freely again!

Rolland's diary gives a unique insight into Einstein's political mind-set. As early as September 1915, Einstein hoped for the Allies' victory. He saw in it the only chance for the destruction of the Prussian dynasty's power and a democratic renewal of Germany, which was otherwise incapable of initiating such a renewal. But, Rolland noted, a German defeat was not foreseeable. "Einstein says that one cannot imagine the organizational power that has been demonstrated and all the capable minds involved in it."[330]

When Einstein returned to Berlin not long afterward, the activities of the New Fatherland League were already under stricter police surveillance. The military officials in charge of the censorship had significantly limited the organization's work in the past months. Memos were confiscated, leaflets banned, and finally travel and publishing bans were imposed on individual members.[331]

Nevertheless, in October 1915, upon request from Berlin's Goethe Society, Einstein wrote an essay titled "My Opinion of the War." He noted there that the ordinary citizen carefully maintained patriotism in his soul as a shrine for "bestial hatred and mass murder, which he obediently takes out in the event of war in order to use them." He himself regarded national identity as a business affair, somewhat akin to life insurance.[332] These and other passages fell victim to censorship.

But even the shorter version that appeared in an elaborately designed "patriotic commemorative album" is a courageous affirmation of pacifism and the idea of a League of Nations.[333] The fact that Einstein's position appeared alongside contributions by high-ranking military officials and politicians, famous poets and intellectuals, indicates that in the Reich's capital he enjoyed the reputation of being the most significant physicist of his time, even beyond his research field.

Undoubtedly the meeting with Rolland had bolstered him. As the New Fatherland League wanted to appoint him to the international

Central Organization for Lasting Peace, Einstein answered that he had "neither the experience nor the abilities for political affairs."[334] Nevertheless, he agreed to actively collaborate—just one month before the completion of his general theory of relativity.

In this chronological telling, we now suddenly find ourselves at the completion of Einstein's achievement of the century. In the fall of 1915, Einstein was suddenly absorbed again in the theory of relativity. The eight weeks that followed his extensive summer vacation in the Rügen Island and Switzerland count as the "most exciting and straining time" of his life.[335]

Einstein found internal contradictions in the chain of reasoning he had been following step by step since the spring of 1913. It is possible he was made aware of this during his stay in Switzerland by his friend and congenial colleague Michele Besso. But why exactly were his equations false? He was now "immensely thrilled," he noted immediately upon his return, but did not believe he could find the error himself. In this matter his own mind was stuck in a rut.[336]

As understandable as his uncertainty was, it did not last for long. Thinking drove him once again to rethinking. Einstein soon returned to working obsessively on revising his own theory. Thus began a new, more dynamic phase of reflection and searching for the field equations of gravity, during which he resorted to ideas he had earlier dismissed.

His road to success can serve as a lesson to all those who fear failure. With his typical boldness, Einstein laid out the possible solution strategy, first in his mind's eye and then in four successive meetings, to the members of the Prussian Academy of Sciences. With his reflections on physics, he took the stage in front of a professional public without any inhibitions, having to present the results of the previous week as a necessary step or as a failed attempt. Each time, he thought he could finally draw aside the Newtonian curtain. The academic public could only marvel at it. However, his own inner critic reported back, skeptical about what he had achieved.

The scientific drama played out for a time in its quirky manner—

until the final act. On November 25, 1915, as Einstein presented to the Academy the most recent version of his general theory of relativity and the year-long musing that led to the until then unknown field equations, he still used, strangely enough, the same formula he had used three years earlier. This time, however, he interpreted it differently. As his biographer Albrecht Fölsing notes: "Had Einstein's theory of gravitation not been surrounded by an aura of being 'probably the greatest scientific discovery ever,' one perhaps would have tried at the same time to present it as a comedy of errors, naturally of the highest level."[337]

Why did he suddenly find in years-old reflections that for which he had been desperately searching for so long? This question has been occupying historians of science until today. They have been repeatedly making attempts to follow the different directions of Einstein's thought and have always been caught in a labyrinth, where it has been difficult to find the solution. Is such an attempt at reconstruction at all valid?

Einstein's thought can be deduced neither from the minutes of the Academy meetings in the fall of 1915, nor from the preserved letters of this period. To get just a vague idea of his greatest scientific achievement, we have to step out of the chronological order of the events. There are many overlapping temporal layers when thinking about one's thinking.

Gravity or Gravitational Field?

Isaac Newton and James Clerk Maxwell were Albert Einstein's great intellectual mentors and his most important "dialogue partners" on the path to a general theory of relativity. Einstein combined essential aspects of their physics theories. Among others, he adopted the concept of "field" from Maxwell's electrodynamics.

As the son of the operator of an "electrical factory," Einstein had carried with him since childhood a few seminal images of physics at play. As a youngster he had followed with the greatest amazement how the compass needle orients itself toward the earth's magnetic field. As an adult, he insisted one should not settle for the view that a magnet affects iron directly through an empty space. Instead, one should imagine that a magnet produces something physically real in the surrounding space, a "magnetic field." This field influences the piece of iron, causing it to become attracted to the magnet. If the magnet moves, then the influence is not directly noticeable in all parts of the surrounding space but is expanding with the speed of light.[338]

What can we envision as a field that is invisible yet has visible impact? Is it just a workaround to facilitate the explanation of how forces are being transmitted? Einstein rejected these claims outright. "The electromagnetic field is no less real for the modern physicist than the chair on which he is sitting."[339] And in what way are magnetic and electrical fields something "physically real"?

Einstein clarified this with an example: a researcher sets an electrically charged ball into rapid oscillations; the rhythmical movement of the electric charge generates an electrical and magnetic field, the influence of which can be measured near the charge and from a bigger distance. The result, according to Einstein, is an electromagnetic wave. "The oscillating charge delivers energy that traverses space at a definite speed."[340]

Such an electromagnetic wave can move from one place to another, for instance from a radio station to the antenna. Today, someone who listens to the radio, makes calls with a mobile phone, or logs into the World Wide Web would hardly doubt the realty of the electromagnetic field. Under certain conditions, when the charge in the above example stops oscillating, electromagnetic waves lead a life of their own. In that case, the produced waves will propagate, having been detached from their source.

Maxwell's theory of electromagnetism is underlined by entirely different ideas than Newton's theory of gravity. To the question "Why does a stone that we drop fall to the ground?" we usually answer as a rule, "Because it is attracted to Earth."[341] This is because in Newtonian physics, bodies are under the direct influence of forces without any temporal delay and over any distance.

In contrast, according to the theory of relativity, no influence unfolds faster than with the speed of light. Therefore, for Einstein, the gravitational influence is also an indirect effect, namely: "Earth produces in its surrounding a gravitational field that influences the stone and produces its falling motion."[342]

In order to ascertain the limited propagation speed of the interaction and expand his reflections on accelerated movement, Einstein wanted to see gravity as analogous with electromagnetism—as a field. Behind such a concept, though, lurked entirely new inquiries. Whether, for instance, there are gravitational waves parallel to electromagnetic waves—a question that has motivated scientists until the discovery of gravitational waves in 2016, which we will discuss later on.

Einstein, though, was initially concerned with seemingly simple, centuries-old connections in physics. The effect of gravity is universal and, in contrast to electricity and magnetism, it always attracts. All bodies tend to cluster together under the influence of gravity. Hence all objects fall to the ground within earth's gravitational field. And it makes no difference whether it is a piece of lead or a plastic ball, regardless of their chemical composition and mass, they all fall at the same speed, if one ignores for a moment the resistance of air. Why does a gravitational field accelerate all bodies equally? Einstein wondered endlessly about this empirical fact and tried to understand its deeper meaning.

He pictured a cart on a smooth plane. According to the law of inertia, if pushed it will move at a constant speed. If the cart is loaded, it will oppose the next push with a stronger inertial force and will roll more slowly. Its speed after the push obviously depends on its mass, which physicists call "inertial mass," since it brakes its movement.

One can measure the inertial mass of the cart. That, however, is laborious. In everyday life, we put an object on the scales. Since weighing with scales is possible only due to earth's gravitational field, in this way we determine the gravitational mass of the body. Here's what's amazing: both methods amount to the same thing. Although the first method has nothing to do with gravity, "inertial mass" and "gravitational mass" are the same.

The significant law of falling bodies is based on this correspondence: when we drop a lead ball, a comparatively strong force is at work between the ball and earth that should lead to a faster acceleration than that of a falling plastic ball. Yet the lead ball has a stronger resistance to a change in motion. The two effects cancel each other out. A lead ball and a plastic ball fall at the same speed.

Classical physics provided no explanation for this odd coincidence that manifests itself in every exact measurement to this day. Einstein ruled out that the equivalence of gravitational and inertial mass was a coincidence. As Einstein noted, "A detective story that arranges mysterious events as accidental is not worth much." Similarly, a theory that would offer an explanation for the fact that gravitational and inertial mass arc identical would be superior to others.[343] He wanted to find this explanation.

Physics in the Elevator

Einstein was a master at posing simple questions and never letting go, seeing problems where, from the perspective of his contemporaries, there were none, and subsequently mapping out the big theories that fell beyond their scope. "He had the gift to see behind inconspicuous facts the meaning that escaped everyone else," his colleague and longtime companion, physicist Max Born said in retrospect. "It was this improbable empathy for the ways in which nature works and not his mathematical abilities that distinguished him from us all."[344]

At some point, as he reflected again on the laws of falling bodies, his imagination was inspired by an image. In the fall of 1907 he had the previously mentioned "happiest thought" of his life: that someone who is free-falling no longer feels any gravitational field.[345]

In the case of a plummeting elevator whose supporting cable has snapped, most of us think only of the dramatic consequences. Not Einstein. It occurred to him that conventional scales would no longer show any weight in this situation. If a horizontal bar were installed in the plummeting elevator, pull-ups with the little finger would suddenly be no problem. But beware! One would have to brake the upward movement so as not to overshoot. Whatever other experiments one might also carry out, they would run as if one were situated a long way from Earth in a gravity-free space. Therefore, the same laws of physics should apply in a free-falling frame of reference as in a system free of forces: the laws of the special theory of gravity.

In Einstein's thought experiment, zero gravity is, however, limited to a small space. Imagine someone in the falling elevator places a ball to his left and one to his right. The balls float, and seen from the outside they move down with the elevator toward earth. On closer observation the fall's acceleration acts in the direction of the center of the earth. Therefore, strictly speaking, the course of the two balls is not parallel. During the free fall they are moving slightly toward each other. Complete zero gravity reigns only in a sufficiently small elevator and only for a short period.[346]

Einstein extended this conclusive theory step by step into a new theory of gravity. Proceeding from his elevator physics, he next envisaged comparable situations—for instance, a completely closed cabin somewhere in space. If objects such as keys or handkerchiefs that are dropped in the cabin fall to the ground, there can be several causes for this. It seems to the occupants as if they are located in the gravitational field of a celestial body. But it may also be that the space capsule is accelerating upward by some power, as with an elevator when it starts to rise. They cannot decide what their situation is. Both views are reasonable from the perspective of physics.

Let us think, for instance, of a spring on the ceiling of a space cabin, from which hangs a weight. If the spring is protracted, then the reason may be a gravitational field pulls it down. But it is equally possible that the inertia of the mass becomes noticeable in opposition to the acceleration of the space cabin as a tension in the opposite direction. Inertia and gravity, acceleration and gravitation, are equivalent.

Einstein called this recognition the "principle of equivalence." It formed the crystallized nuclei of his new theory of gravitation. "The heuristic value of this assumption lies in that it allows us to replace an homogenous gravitational field with a uniformly accelerated frame of reference, the latter being amenable to theoretical examination up to a certain degree," he wrote at the beginning of his studies in 1907.[347]

Einstein remained faithful not only to the question he had once raised, but also to the physical images and thought experiments he hoped would provide a clarification. The box and the cabin in space always reappear in his mindscape. Little did he know that a hundred years later, around one thousand "boxes" would revolve in the near-earth space, artificial satellites deployed for communication purposes, meteorology, earth observation, and navigation. The general theory of relativity acquired a practical importance. Einstein, however, with his thought experiments, was aiming at finding a new order for natural phenomena.

Bearing this fact in mind, we return to the cabin. Next, we imagine a source of light on the interior wall. It sends a light ray horizontally into the room. As long as the cabin, seen from the outside, moves with constant speed, the light ray strikes exactly opposite the point of incidence. This changes, though, when the space capsule constantly accelerates upward. In this case the light ray bends downward. It will draw a slightly curved course.

Yet a passenger would interpret this matter differently. It would appear to her as though the cabin is at rest but within the range of a gravitational field. When the light ray is diverted from its otherwise straight course, the cause should lie with a gravitational field.

Entirely confident in the "principle of equivalence," in 1907, namely eight years before the completion of his general theory of relativity, Einstein predicted that light deviates from its course in a gravitational field.[348] He calculated that for a ray passing directly by the sun, the angular deflection would amount to just 0.83 arcseconds: not even as much as the diameter of a pea, observed from a distance of one kilometer. Einstein admitted that the assumption on which his calculations were based "even if obvious, are nevertheless quite bold."[349] Still, he was hopeful that astronomers would verify his prediction during a total solar eclipse, when the moon slides in front of the solar disk. For a short period of time, those stars in the immediate vicinity of the sun that are otherwise outshined by its light, would be visible. During an eclipse, one can photograph the position of those fixed stars in the solar vicinity and subsequently compare the results with photos of the same group of stars taken at night when the starlight does not pass by the sun on its way to earth.

The astronomer Erwin Freundlich was excited with this idea. Yet his long prepared solar eclipse expedition to Crimea failed. As already mentioned, he and his team were briefly taken as prisoners of war by the Russians. Back in Berlin, Freundlich hatched new plans and received, at least temporarily, support from renowned physicists, among them Arnold Sommerfeld, who appealed to his colleagues to test Einstein's theory as soon as possible. "Make sure that German astronomy does not make a fool of itself! It hasn't had an opportunity like this for decades to demonstrate its high level."[350]

The Discovery of Slowness

In Munich Sommerfeld studied the structure of atoms and tried to understand why atoms emit light only at certain wavelengths. Consequently, he was interested in those aspects of the general theory of relativity that concerned light. His pen friend Einstein came up with further surprises in this respect. Let us return to the space capsule.

This time there are two physicists on board, let us call them Einstein and Planck. They are provided with identically constructed instruments. Planck, on the ground of the cabin, sends light waves upward toward the ceiling, where Einstein holds out a detector. After

Einstein receives the light signals, he answers with the corresponding light pulses that he sends out toward Planck.

We have already recognized light as an electromagnetic wave. The sequence of wave crests and troughs is characteristic of those light waves. That is how the rainbow colors, of which the light spectrum of the sun is composed, differ from each other, as the distance between the waves' crest of violet, blue, green, and yellow up to red is constantly increasing. What electromagnetic waves do Einstein and Planck perceive on our space capsule?

When the cabin constantly accelerates upward, Einstein, at the top of the space capsule, moves with ever increasing speed away from the light waves that reach him from below. The temporal distance between the wave crests increases. Consequently, the light's color changes as well. It shifts toward the red end of the spectrum.

Planck observes the opposite effect. On the ground of the cabin he heads always faster toward the place from which Einstein sends his light signals. The wave crests are coming toward him in close succession. From his perspective, the color shifts to the blue end of the spectrum. We encounter something similar in everyday life with sound waves, namely when the police or the fire brigade drive by and the sound of the siren changes as they pass—the wavelength, that is the pitch, changes.

We can go back now to the "principle of equivalence." Einstein and Planck have the impression that their cabin is at rest, located in a gravitational field somewhere on the surface of a celestial body. Their conclusion: the gravitational field causes the light that distances itself from the celestial body to appear red. This color change—physicists speak of "gravitational redshift"—ought to occur especially on the sun.

Einstein chose distinctive colors from the light spectrum, such as the yellow light of unified wavelength emitted by sodium atoms into the solar atmosphere. When it propagates away from the solar gravitational field, the light should arrive on Earth with mild redness. In his 1911 article "On the Influence of Gravitation on the Propagation of Light," he was skeptical as regards the question of whether the negligent shift of the spectral line will be susceptible to observation.[351]

The director of thc observatory in Potsdam, Karl Schwarzschild, was more optimistic in this respect. He did not think such a test would be hopeless and installed a spectrometer for observing sunlight in the tower of a residential house for civil servants. But the instrument did not have the sufficient spectral resolution for proving the redshift.[352] A better instrument was necessary, for which the financial means were not available, due to the war. Furthermore, Schwarzschild signed up voluntarily for military service and from then onward calculated the trajectory of shots.

Once again Einstein had to exercise patience. Undeterred he held on to the implications of the "principle of equivalence." Light as electromagnetic waves continued to play a key role in his studies of gravitation. Einstein considered light waves as one of many periodic phenomena. It was the visible expression of a rhythmical atomic process. Light generating atoms presented for him miniscule clocks, the "running of which is indicated by the wavelength of the radiated light."[353]

With this image in our head, we return once again to the situation described above, inside our space capsule where light signals run from top to bottom and in the opposite direction. Now we can interpret the obtained results differently. In comparison to Einstein's clock, which is situated higher, Planck's clock, on the ground of the space capsule, runs slower. Time dilates under the influence of gravity.

This sounds fantastic. Looking down from a high-rise at what is happening on the street, it does not appear as if the people down there are living with a slower rhythm. Indeed, Einstein's calculated gravitational time dilation on Earth's gravitational field is small. So small, that there was no prospect during Einstein's time of measuring it with clocks.

In the meantime, Einstein's prediction has been often verified. Owing to the extreme accuracy of today's atomic clocks, it has sufficed to leave one clock on the ground and heave a second, identically constructed atomic clock up to a higher floor, or even just put it on a table, in order to recognize that the clock on the ground ticks slower. For instance, in an experiment in Boulder, Colorado, in 2010, researchers ascertained that the oscillations of two atomic clocks at a height difference of 33 centimeters differ by a factor of 0.000000000000000041.[354]

You may feel disappointed: so much fuss about such a minute amount! But as we will still see, the time delay in strong gravitational fields in the universe is significantly bigger. In extreme cases time can come to a standstill.

Above all, the comparison between two atomic clocks makes us aware of the modern research laboratory, of how closely time, space, and matter are interlinked. A modern atom clock is a much more reliable timer only when scientists take into account the height at which the clock is situated in relation to a specific reference surface in Earth's gravitational field. Conversely, precise atomic clocks can assist researchers to determine small height differences. Researchers are also able to measure earth's gravitational field with the same atomic clock—again, under certain theoretical assumptions and experimental conditions. Measurement of time, length, and gravity are mutually dependent.

In Newtonian physics, space and time are absolute realities. They form a container of sorts for all physical occurrences. All processes take place in absolute space and time, which remain unaffected by any incidences.

Einstein corrected this image in his special theory of relativity by expounding on the reason why space and time cannot be separated from each other. Since simultaneity has no absolute meaning, space and time coalesce into a spacetime continuum. In systems moving quickly against each other, the scales of time and length differ.

In contrast, according to the general theory of relativity we are no longer concerned with a fixed spatiotemporal regulatory framework. Together with matter, space and time form a dynamic framework. In it, gravity is a component of the spacetime structure.

Today, the slight gravitational time dilation reaches our everyday life, most notably in navigations with GPS devices. In order to determine the actual position on Earth, the navigation device establishes contact with numerous satellites that orbit around the planet. The distances to the satellites are eventually determined by the duration of the signal. But since the satellites are circling far above us where time, loosely speaking, does not elapse as slowly as down here on Earth, the influence of earth's gravitational field has to be constantly taken into consideration in order to limit mistakes in navigation.

For the sake of completeness, it should also be noted that in satellite-based navigation, the special theory of relativity cannot be disregarded. Consequently, for the GPS satellites, clocks are moving relative to the ground. From the perspective on Earth, they run more slowly. The difference that results from this relativistic movement is significantly smaller, though, than the one owing to Earth's gravitational field as described above.

Curved Spacetime

We became acquainted in this chapter with a whole range of physical phenomena that Einstein derived from the "principle of equivalence" without using much mathematics: time runs more slowly near a heavy-mass celestial body than in a farther distance; under the influence of a gravitational field, a light ray changes its color and continues further with a curved course.

We have discussed elsewhere the deflection of light as that of the sun. The word "deflection" may illustrate the fact but veils the essential aspect of the general theory of relativity: namely that gravitation determines the geometrical structure of space and time and thereby also what is to be understood as "deflection" or a "straight" course. But even that is not certain.

Isaac Newton drew on the familiar concept of a "straight line" in order to define his fundamental laws of physics: a body unaffected by external forces moves uniformly in a straight line. The force-free inertial motion takes place along a straight line. In comparison, Newton treated gravity as an external-force effect—namely, gravity knocks a thrown ball off its otherwise straight course by bending its trajectory into a parabola. The same force also binds the moon to Earth and causes planets to travel around the sun in an elliptical orbit.

Einstein, in contrast, did not look at gravitation as an external force. Just as gravity is suspended in a free-falling elevator, a body also moves free of any force in a gravitational field. In accordance with the Einsteinian theory, the body's movement occurs through inertia *and* gravitation as linearly and uniformly as possible. These straightest possible world lines, called "geodesics," play a similar role in the theory of relativity as straight lines do in classical physics.*

Naturally, in order to remain within our trusted Euclidean geometry, one could nonetheless say that a light ray is "diverted" in a gravitational field. Einstein was determined, for reasons of physics, to incorporate gravitation into the geometrical structure of space and time. He wanted to find natural laws that would have the simplest possible form for all observers, even for those who are accelerating toward each other. He was able to do this only by orientating himself toward a geometry of physics based on the distribution of matter. Consequently, he assumed that in a strong gravitational field, geometry of space would no longer correspond with our conventional notions—in other words, it would no longer be Euclidean. In particular, he considered the light-ray path as a general straight line, as "geodesic."

What does this mean? Let us take a look at the curved surface of the earth. We can find geodesics on the globe's surface without

* This can be elucidated with the example of light rays. Until Einstein's prediction of light bending by a gravitational field, researchers assumed that in an empty space, light rays constantly propagated in a straight line. In geodesy, light rays are the means of choice for defining "straight" routes and measure the distance from A to B.

When you invite a painter to your house these days, he usually has a handy laser meter that shows the actual distance between the walls to a millimeter.

much effort. We select any two points—for example, Berlin and New York—and connect them with a path a plane would chose. If we follow this straightest possible line with a finger time and again around the globe, we eventually arrive at the starting point. On a spherical surface, the "geodesic" is always part of a big circle.

The longitudinal meridians that run from the north pole to the south pole are also geodesic lines. They cut the equator at a right angle. If we draw a triangle from the north pole to the equator, further along the equator and back over the longitude to the north pole, then the sum of angles of this triangle will be bigger than 180 degrees. This already indicates we are dealing with a curved surface, since on a flat plane the sum of the angles in a triangle is always exactly 180 degrees.

Generally, we do not witness this curving of the earth's surface. If we sit in a car and pull away, we are not under the impression that we are moving on a spherical surface. Seen from a local point of view the earth is flat. Therefore, one can draw a map of a small section of Earth's surface using the same scale for the entire region. On a true-to-scale map of the Berlin surroundings, for instance, one centimeter corresponds to four kilometers in reality.

By contrast, Earth's surface cannot be represented on paper true to scale in its entirety. Presumably you have seen world maps that represent the entire globe. On them the polar regions are usually quite wide. For such maps it applies that the further one moves south or north from the equator, the smaller is the actual route that corresponds to one centimeter on the map.

If you would like to discern the actual relations on such a map, you can no longer revert to a fixed scale. Instead, you have to know the corresponding mathematical context for each point on the world map. The mathematician says: you have to know the metric. Only the metric enables us to calculate the actual distance between two random points. Curved surfaces require a variable metric. The metric in the map of Berlin and its surroundings is constant.

In Newton's classic theory too, the metric is fixed. Space is an absolute, its geometrical structure and scales are equal everywhere.

Therefore, the "geodesics" of classical physics are the familiar straight lines, and the sum of angles in a three-light-rays spanning triangle will always be 180 degrees.

In his special theory of relativity, Einstein dealt with four dimensions: three spatial directions and time. For Einstein, space and time formed a unity. In the special theory of relativity he thought that the simplest way to describe physical phenomena is with a four-dimensional spacetime that has a fixed metric. That changed, though, as soon as he began to include gravity in his reflections.

Suddenly he was confronted with the bending of light rays, gravitational time dilation, and other surprising physical effects. That he joined all this to a new theory of gravity demonstrates the unique depth of his mind. His bold conclusion: a rational ordering of space and time ensues from the to-be-organized bodies themselves. It is the existing masses that prescribe the metric of four-dimensional spacetime.

The division of matter and energy goes hand in hand with a gravitational field that determines how geodesics look. Light rays expand along these straightest possible lines. But planets too, move along geodesics. They behave similarly to balls that roll over an uneven surface and are diverted into a curve as soon as they roll into a hollow. In this envisaged surface, the sun generates a special spacious recess, in which the planet-balls are caught; their course is predetermined by the sun's gravitational field.

To us, the orbits of the planets, asteroids, and comets do not look as if they constitute the straightest possible route. But we are unable to imagine their route as curves in a bent four-dimensional space. Our entire perception is bound to a three-dimensional world. As a result, for centuries, even mathematicians considered it senseless to be dedicated to a geometry that involved a potential fourth or fifth dimension.

It was first in the nineteenth century that an intensive debate began about n-dimensional geometries, to which Einstein could refer when he wanted to describe the movement of bodies along a curved surface. But he had to build the bridge from Euclidean to non-Euclidean geometry by himself. Einstein's friend, the Nobel Prize laureate in physics Max von Laue, outlined his position: "This required mathematical tools, which no physicist had ever used, and which led a rather concealed existence even in mathematical literature."[355]

In the Thicket of Mathematics

How to proceed then? In desperation, in 1912 Einstein turned to a former fellow student, who in the meantime had become a mathematics professor in Zurich. "Grossmann, you've got to help me, or I will go crazy!"[356] Grossmann was no expert in curved spaces and surfaces, but he knew the mathematical tools that had been developed in the nineteenth century. More importantly, Grossmann took all the time necessary to familiarize both himself and Einstein with modern geometry.

From that point onward Einstein's high esteem for mathematics grew from week to week. He let his physicist colleagues know that never in his life had he been as frustrated as with the geometrization of gravity. "Compared with this problem, the original theory of relativity is mere child's play."[357]

Most difficult for him was calculating the gravitational field from the available matter and energy. He oriented himself toward Maxwell's theory. But how should he derive the field from a four-dimensional variable metric that describes the bending of space? He had to deal with an operand of ten independent components. Einstein had to laboriously learn how to use such complex mathematical objects.

His notes from this period refer to exercises that are all too willingly omitted in favor of "inspiration" and "sudden intuition." Einstein's masterwork is based on continual exercises. The repetitive character of his work manifests itself today in mathematics all over, for in this respect Einstein depended on paper more than on his physical observations. For exercise purposes, in addition to notebooks he scribbled on sandwich paper and on his cuffs.

To build the necessary framework for his ideas in physics, he acquired mathematical tools, which he enriched incrementally. Searching for general field equations, he next dedicated himself to simple problems such as weak and homogenous gravitational fields. Time and again he returned to special cases for which classical physics was useful and for which he believed he had already known the solution.

As Einstein's notebooks reveal, his cooperation with Grossmann bore fast fruits. The research duo jumped big thematic hurdles within months and had the correct solution in view. "Two years before the publication of the general theory of relativity we had already considered the right field equations of gravitation," Einstein said in retrospect. "But we were unable to recognize their usefulness for physics."[358] The required equations were already on paper, but the physicist did not identify them as such and scrapped them again.

Other exceptional theoreticians had not fared better. Three hundred years before him, Einstein's famous predecessor Johannes Kepler had found the elliptical equation of Mars's orbit but could start nothing with it. Kepler jumped back and forth between mathematics and physics, but the full meaning of the formula did not open up to him, as long as he was trapped in old thinking patterns. Only after months did he become aware of his blindness: "Oh, what a foolish coot I am!"[359]

Einstein had a similar sense of self-irony. He too laughed at his own mistakes. Like Kepler, he was strictly committed to observations and measurement results but at the same time free to choose his research methods. Both scientists approached the same phenomena with always new approaches from different angles and in light of new insights. Kepler's departure from the geometry of the circle was equally unprecedented in the history of astronomy as Einstein's later break with Euclidean geometry.

It was not apparent, even in the simplest cases, how Einstein could give his mathematical calculations a physical interpretation. Hence, he thought that the field equation of the general theory of relativity with its ten components, had to, when transitioning to Newtonian theory, transform into an equation with just one single component. Since this did not happen, he considered the right solution to be wrong.

Instead, in 1913 he and Grossmann drafted a different calculation that led in the hoped manner to Newtonian physics. But contrary to his generalized principle of relativity, these field equations now no longer had the same form for all observers. He found this embarrassing. But there were possibly physical reasons for the qualified validity of the principle of relativity. For the time being, Einstein got out of the affair on the grounds that the new theory of gravitation had to satisfy a few primary, limiting principles of physics, for instance the law of conservation of energy or the principle of causality. It cost him another two years of reflection to correct his notions in this respect and return to his general principle of relativity.

Mathematics or Reality

During these two years he observed his *Entwurf* theory (the draft) more or less skeptically, spoke frankly and openly about intellectual hurdles and progress, and published a few works on the physics of gravitation. Inevitably, other researchers jumped on the wagon, among them David Hilbert. In the summer of 1915 the fifty-three-year-old mathematics professor invited Einstein to Göttingen for a lecture series, and a few months later emerged as a contestant for the completion of the general theory of relativity.

Hilbert was one of the most eminent contemporary mathematicians, having become famous among other things through his list of twenty-three "house assignments" presented to his colleagues as mathematical problems for the twentieth century. He himself searched for an axiomatic foundation of mathematics. The meaning of this can be explained with the following example from geometry.

For thousands of years, mathematicians assumed that all geometrical theorems could be traced back to a few postulates. Everything else could be derived through logical conclusions from these axioms, which Euclid set at the beginning of his geometry. Although the validity of the axioms could not be proven, they counted as evident or, in Hilbert's words, "as extracted from complexes of experience."[360]

Only in the nineteenth century did it become apparent that it was possible to vary these axioms and thus arrive at other consistent theories of non-Euclidean geometry. Hilbert considered such geometries just as "true" as Euclidean geometry. He regarded mathematics as a purely formal game with symbols and a set of rules. Instead of referring to evidence of postulate axioms, he approached them as hypotheses, in order to then prove whether they led to a consistent theory. Hilbert had already produced the formal proof for the consistency of Euclidean geometry.

Einstein paid respect to Hilbert's mathematical research. The obtained progress was the neat separation of formal logic from descriptive content. According to axiomatics, formal logic is the only object of mathematics; it is based on thinking, not experience. "But such a pure exposition makes it evident that mathematics cannot predicate anything about objects of descriptive presentation nor about objects of reality," Einstein declared emphatically.[361] Stated more clearly: "As far as the laws of mathematics refer to reality, they are not certain, and as far as they are certain, they do not refer to reality."[362]

However, the very words "geometry" and "geodesy" indicate that particular structures in the physical world led mankind to mathematical thinking. Einstein discerned a branch in the historical development of mathematics that he called "practical geometry." It deals with the disposition of natural objects in respect to one another, for instance a measuring rod in relation to parts of Earth, and assigns mathematical concepts to objects of experience. All measurements of length in physics or astronomy are attributed to this kind of geometry.

Seen from this point of view, geometry and physics are mutually dependent. Experience teaches that there are no rigid, unchanging objects. Each measuring rod can be deformed and changes its length when the temperature rises. Every correct use of a measuring rod requires knowledge of physics that is itself based on geometry. The relation between geometry and experience is thus revealed only in the framework of comprehensive reflection. In this context, according to Einstein, the question whether "practical geometry" of the world is Euclidean or not gained a distinctive meaning.[363]

The tension between basically precarious knowledge based on experience and deductively gained knowledge of mathematical axioms shaped Einstein's relation to Hilbert. The Göttingen mathematician wanted to transfer his axiomatic method into physics. "I believe everything that can be the subject of scientific thinking becomes, as soon as it is ripe enough to constitute a theory, subject to the scope of axiomatic method and thereby mathematics."[364]

Hilbert considered physics too challenging to be left in the hands of physicists alone. For a number of years, he had been dealing with actual research themes in physics that seemed to him "ripe" for an axiomatic approach. He hired physics tutors to read to him and discuss the latest articles on solid-state physics or radiation theories. Furthermore, the Göttingen mathematician invited luminaries in the field—from Planck to Sommerfeld—to give guest lectures. In the long run, he wanted to base physics, like mathematics, on a secure foundation of reasoning.

A sample of his ability came in 1912, when he deduced a physical radiation law out of a few specific basic assumptions. Even non-mathematicians marveled at his solution strategy. It included, however, neither any reference to experiments nor a plausible explanation why this law should be applicable. The criticism of his approach from expert physicists was foreseeable. The general tenor: Hilbert's axioms already included what needed to be demonstrated beforehand.[365]

As Einstein arrived in Göttingen, he could not have known that his opponent had a similar idea regarding the theory of gravitation: to deduce it from few axioms. Einstein spent the week of June 29, 1915, in Göttingen at the Gebhards Hotel, and gave six two-hour lectures at the university. He based his lectures above all on his paper published in the autumn of 1914, "The Formal Foundation of the General Theory of Relativity," which had already been the subject of discussion in Göttingen.[366] Einstein felt he was understood by the scholars in every detail, especially by the mathematician, who liked to conceal his bald head under a Panama hat. "I am completely thrilled with Hilbert," he raved. "An important man!"[367]

Einstein's appreciation was not only for Hilbert the mathematician, but also Hilbert the liberal political thinker. Admittedly, Hilbert got carried away at the war's onset by the general war enthusiasm and signed, along with three thousand professors, the nationalistic "Declaration of Professors of the German Reich."[368] Since then, though, his attitude toward the war had changed, as shown in the debate over the objectives of the war, which reached its new climax in the summer of 1915.

While Einstein was in Göttingen, the Berlin theologian Reinhold Seeberg was at the universities, promoting support for German expansion politics. With him, 1,347 signatories, among them 352 German university teachers, demanded that in case of a peace settlement Germany should keep Belgium, the French Channel coast, and the Baltic States in the east. The historian Hans Delbrück particularly opposed the annexation to the west, which was no longer part of a war of defense, and composed a moderate memorandum. It found only 141 proponents from various political positions, among them Einstein, Planck, and Hilbert.[369]

In the close scientific circle in Göttingen, Hilbert created around himself a politically free group that provided colleagues with some protection. Einstein spoke openly with his colleagues about his own pacifist commitment in the New Fatherland League and thereby captured the interest of the mathematician and theoretical physicist Paul Hertz, whom he had previously met in Switzerland.[370] Their subsequent correspondence clarifies what vexed Einstein during the war years.

Between August and October, the two physicists corresponded

about possible ways of overcoming a few difficulties in the existing theory of gravity. Along the way, Hertz got caught up with the activities of the League and couldn't make up his mind whether he should join it. He considered obtaining its regular announcements through a front man. After some back and forth, Einstein advised the worrywart against a membership: "You have that valiant attitude that those in power greatly love in Germans." Fine people the likes of him provided the best guarantee for maintaining the political swamp. "Be assured that I love your mind, even though I pity you for being spineless."[371]

Half a year earlier Einstein would not likely have written such a strident letter. This document alludes to possibly similar controversies with Berliner scientists. Considering Einstein's temperament, it is likely that he occasionally "unmuzzled" himself in the capital as well. Einstein traced the main reason for Hertz's and other Germans' inability to escape public opinion to the education system. Everyone acted in accordance with what had become to some extent their conditioning, and regarded themselves as good, useful, and irreplaceable.[372]

Upon receiving this letter Hertz broke off contact, although Einstein immediately apologized for his rudeness and assured him that he would maintain his friendly attitude in the future. Seven years later, when Hertz, who was of Jewish descent, came into financial difficulties, Einstein prepared a letter of recommendation for him.[373] One of the many examples of his tolerance and humanity, for all his intellectual sharpness.

The rift with Hertz, who shortly afterward was drafted to the Imperial German Army Air Service, clearly did not impair Einstein's relationship with Hilbert. However, Hilbert had lost in Hertz an engaged interpreter of the new physics of gravity and of all its phases, precisely at a time when he was more than ever immersed in Einstein's physics. He was not surprised that Einstein had taken steps toward establishing a theory based on non-Euclidean geometry. For why should that which applies for technical instruments not apply for mathematical tools: namely that the range of possible applications grows ever bigger than the specific aim for which it had been devised. Physics should weave modern geometry into its investigations. According to Hilbert, "Each science grows like a tree; not only do the branches reach further out, but the roots, too, reach deeper."[374]

The main ideas of the general theory of relativity immediately made sense to the mathematician, such as the independence of physical propositions from the chosen system of coordinates.[375] Only its structure seemed to Hilbert inconvenient and unnecessarily complicated. In the last part of Einstein's "Formal Foundation of the General Theory of Relativity," Hilbert hit upon a process that became his own key for revising the entire theory: the so-called "Lagrangian formalism."* Hilbert put it at the beginning of his own work, and thereby upended the Einsteinian approach.

However, the mathematician intended to build the physics of gravity not just axiomatically. Moreover, he recognized structural similarities between Einstein's theory and a new theory of matter that referred back to the German physicist Gustav Mie. This led him to the original thought of uniting the two under the umbrella of a "*Weltfunktion*" (world function).

"The Most Valuable Discovery of My Entire Life"

While Hilbert started searching for the *Weltfunktion* in the summer of 1915, Einstein went on vacation. He spent the first half of his summer holidays with Elsa and her daughters on the island of Rügen. The holiday by the sea with his "small harem" made him happy. He had just recently raved to his friend Besso about his extremely beneficial, beautiful relationship with his cousin. The endurance of their

* Lagrangian formalism is one of the main tools for describing the dynamics of a vast variety of physical systems, including systems with a finite (as with particles) and infinite (as with strings, membranes, fields) number of degrees of freedom.

Die Grundlage der allgemeinen Relativitätstheorie

~~A. Prinzipielle Erwägungen zum Postulat der Relativität.~~

~~§1. Die spezielle Relativitätstheorie.~~

Die im Nachfolgenden dargelegte Theorie bildet die denkbar weitgehendste Verallgemeinerung der heute allgemein als „Relativitätstheorie" bezeichneten Theorie; die letztere nenne ich im Folgenden zur Unterscheidung von der ersteren „spezielle Relativitätstheorie" und setze sie als bekannt voraus. Die Verallgemeinerung der Relativitätstheorie wurde sehr erleichtert durch die Gestalt, welche der speziellen Relativitätstheorie durch Minkowski gegeben wurde, welcher Mathematiker zuerst die formale Gleichwertigkeit der räumlichen Koordinaten und der Zeitkoordinate klar erkannte und für den Aufbau der Theorie nutzbar machte. Die für die allgemeine Relativitätstheorie nötigen mathematischen Hilfsmittel lagen fertig bereit in dem „absoluten Differentialkalkül", welcher auf den Forschungen von Gauss, Riemann und Christoffel über nichteuklidische Mannigfaltigkeiten ruht und von Ricci und Levi-Civita in ein System gebracht und bereits auf Probleme der theoretischen Physik angewendet wurde. Ich habe im Abschnitt B der vorliegenden Abhandlung alle für uns nötigen, bei dem Physiker nicht als bekannt vorauszusetzenden mathematischen Hilfsmittel in möglichst einfacher und durchsichtiger Weise entwickelt, sodass ein Studium mathematischer Literatur für das Verständnis der vorliegenden Abhandlung nicht erforderlich ist. Endlich sei an dieser Stelle dankbar meines Freundes, des Mathematikers Grossmann gedacht, der mir durch seine Hilfe nicht nur das Studium der einschlägigen mathematischen Literatur ersparte, sondern mich auch beim Suchen nach den Feldgleichungen der Gravitation unterstützte.

A. Prinzipielle Erwägungen zum Postulat der Relativität.

§1. Bemerkungen zu der speziellen Relativitätstheorie.

Der speziellen Relativitätstheorie liegt folgendes Postulat zugrunde, welchem auch durch die Galilei-Newton'sche Mechanik Genüge geleistet wird: Wird ein Koordinatensystem K so gewählt, dass in Bezug auf dasselbe die physikalischen Gesetze in ihrer einfachsten Form gelten, so gelten dieselben Gesetze auch in Bezug auf jedes andere Koordinatensystem K', das relativ zu K in gleichförmiger Translationsbewegung begriffen ist. Dies Postulat nennen wir „spezielles Relativitätsprinzip". Durch das Wort „speziell" soll angedeutet werden, dass das Prinzip auf den

Image 7: *The beginning of Einstein's manuscript of the first summary of the general theory of relativity in the Annals of Physics in May 1916.*

relationship could only be guaranteed by renouncing his marriage.[376]

The subsequent trip to Switzerland assured above all a reunion with his children after more than a year. Einstein had written enthusiastically to his older son beforehand about a summer trip being "fourteen days or three weeks alone just with you." Perhaps even hiking in Italy. "I will also tell you many beautiful and interesting stories about science and other things."[377]

This announcement caused no particular joy in Zurich. Instead, it initiated bickering between the parents. Einstein accused Mileva of turning the children against him. Vengeful as she was, she presumably dictated to her son the line "as long as you are not friendlier with mama, I will not come with you…."[378] In the meantime, Einstein no longer received a response when he wrote to Hans Albert.

The letters Einstein had been writing to his wife since the separation were cold, reproachful, mean, then amicable, attempting to simplify the situation. The letters, often about financial questions, dealt mostly with his contact with the children. Only one of Mileva's letters from 1915 has been preserved; however, Einstein's correspondence with Michele Besso and Heinrich Zangger gives a sense of how especially difficult the separation was for the two sons. Besso and Zangger repeatedly reproached Einstein for lacking empathy.

When in September 1915 Einstein arrived in Zurich, he became insecure; his sons avoided him. He saw them just twice. "Thereafter stalemate. The mother fears the little ones will become too dependent on me."[379] And what of his own fear? Don't his rare letters to his children also prove how inconstant and superficial his fatherly feelings were?

On the other hand, reuniting with his old friends did him good, just as meeting with the pacifist Rolland. The French writer called on European academics to not just stare rigidly at the mindless slaughter at the front. "Like Archimedes, pursue your work in the besieged city. Don't you think a scholar who solves a problem in the midst of a human thunderstorm is more useful to the world than the Manifesto of the Ninety-three Intellectuals? Let us work on the eternal things….

Just as the Red Cross dresses bodily wounds in battle, we have to hurry to help save the spirit."[380]

Immediately upon his return to Berlin, Einstein fell into a creative frenzy. Once his attention had been drawn to the flaws in his present theory of gravity, he initially had little hope he was able to put it right. Only a "fellow human being with unspoiled brain matter" could perhaps accomplish that.[381] Shortly afterward he found out that this "fellow human being" in Göttingen was eagerly focusing on the matter. The thought that Hilbert might have uncovered a fly in the ointment left Einstein no peace. The mathematician had all of a sudden set a pace that Einstein himself could hardly keep up with.

Once Einstein took up the old threads again, he spent day and night calculating feverishly every equation, from which "I parted, though with a heavy heart, three years ago, when I worked together with my friend [Marcel] Grossmann."[382] Forgetting to eat and sleep, he went once again with tremendous energy and intensity through all the possible calculation options in order to finally figure out in what way the presence of matter and energy determines the structure of space and time. In the past three years he had handled all kinds of examples. This wealth of experience opened for him in the fall of 1915 a new leeway for interpretation.

On November 4, 1915, he submitted to the Prussian Academy of Sciences the reputed solution of his problem. First, he admitted that he had completely lost trust in his previously constructed field equations.[383] Subsequently he praised his new work. "The magic of this theory will hardly go unnoticed by anyone who has truly grasped it; it signifies a true victory...of the general differential calculus."[384] Curious to know Hilbert's assessment, he immediately sent his calculations to Göttingen.

In the following week Einstein appeared before the Academy again, this time with a supplement he had founded on an "even stricter logical construction of his theory."[385] He informed Hilbert about it as well, receiving from him in return an invitation to come to Göttingen immediately. Hilbert announced he would present his own axiomatic solution on November 16. "I consider it ideally beautiful, mathematically." And absolutely compelling. On Einstein's paper from November 4, he just commented briefly that it was "entirely different" from his own solution, especially since he, Hilbert, dealt with gravity and electrodynamics together.[386]

Einstein was now really curious. He wrote in return that Hilbert's field equations interested him "extremely," as he himself racked his brain, trying to find the bridge between gravity and electrodynamics.[387] Nevertheless he declined the short-notice invitation to Göttingen on health grounds. He was plagued, as often happened, by stomach pains.

Besides, he was overtired. His adventurous calculations led him deep into astronomy. Unerring, he singled out one aspect of his theory, which he thought was the key. On November 18 he had something essentially new to report to the Prussian Academy. Given the significance of his results, this time he presented them orally.

It was about the planet Mercury, the innermost in the solar system. For a moment in the fall of 1915 the small, rather inconspicuous planet became the most important touchstone of the general theory of relativity. Had Mercury been subject to the sun's gravity only, according to Newtonian physics it had to move in an elliptical trajectory and return to its starting point after each orbit. Yet Mercury's orbit is by no means a closed ellipsis. The planet's orbit around the sun forms a complicated rosette-like trajectory.

That alone would have provided no grounds for doubting classical physics. Newton had already realized that a precise comparison of a theory with astronomical observations has to take into consideration not just the sun's gravity, but also that of the neighboring planets. First of all, it is the planets Venus and Jupiter that influence Mercury's orbit. But even when taking in account all arithmetical laws, the observation results still showed a discrepancy with Newtonian equations of Mercury's orbit. According to Einstein, it amounted to around 45 arcseconds per century.

Wherein lay this difference? Is the sun perhaps not completely round and its gravitational field consequently asymmetrical? Is there another, till then undetected planet in the vicinity of Mercury, which orbits the sun? Many astronomers in the nineteenth century looked for this hypothetical celestial body. Driven by the joy of discovery, they traveled from one solar eclipse to the next in order to find the planet "Vulcan." To no avail.

Einstein looked for an explanation on an entirely different level. Since 1907 he had nourished the hope of finding the cause for the peculiar Mercurian orbit in a new theory of gravitation. If it was at all possible to determine deviations from Newton's theory in the orbits of the solar system, it would be for the innermost planet, Mercury.

He dared to resume the tricky calculations with the new field equations from November 11. The explanation gap closed up at a stroke. Within the scope of the new theory of gravity, the planet ellipses revolved in a way that allowed Einstein, at the end of his calculations, to obtain rather precisely the missing part: 43 arcseconds per century. For a few days he was stunned with excitement. For the first time, his own theory of Newton's physics has been proven superior on empirical grounds as well. On November 18 he spoke to the Academy about a complete accordance between the theory and experience and a significant confirmation of "this radical theory of relativity."[388] The very next day, Hilbert congratulated him on his Mercury calculations and said he hoped Einstein would continue to keep him informed.[389]

Einstein must have received an additional letter from Göttingen prior to this. We can only speculate about its content. Presumably, immediately after his own lecture on November 16, Hilbert sent the manuscript or something similar, for Einstein answered him: "The system you presented corresponds—as far as I know—exactly with what I found in recent weeks and presented to the Academy."[390]

What correspondence is he referring to here? Einstein's formulation poses a riddle, as he himself, at this point in time, still has not reached his goal. Despite his sensational explanation for Mercury's rosette-like orbital pattern, he only submitted his final field equations to the Academy on November 25.

And Hilbert? The mathematician submitted his "fundamental equations of physics" for publication on November 20 in Göttingen. The version of this paper that was printed later contained the correct gravitational equations. Had Hilbert beaten Einstein to the answer by five days? Did Einstein actually adopt Hilbert's correct field equations? After all, the two researchers were in close contact with each other through various channels of communication during these hectic November weeks.

At least for the latter there is no convincing proof. Einstein built his field equations step by step with the mathematical building blocks he already knew. In the first version from November 4, 1915, the solution was still incomplete. Just one week later, Einstein calculated with an almost complete field equation. Fourteen days later he completed it with the last missing part, and in fact in the exact manner as three years before.[391] Although the determining act of creativity eludes direct observation, Einstein's last step of solving the gravitation problem follows its own inner logic.[392]

Historians of science, including Jürgen Renn from the Max Planck Institute for the History of Science in Berlin, were able to retrace Einstein's physical considerations and mathematical procedure on the basis of his earlier notes. Notes in his "Zurich Notebook" reveal not just that on November 25, 1915, he had the same equations he had already reached in the winter of 1912–13; they also suggest he added the still missing mathematical term to his equations in order to account, among other things, for the conservation of energy.

This additional component did not change anything in regard to the sensational result concerning Mercury's elliptical turns; hence, the calculation pleased him enormously. Unlike in the winter of 1912–13, he now recognized the physical content of the equations. He just had to correct an earlier result: the deflection value of a light ray passing close to the sun was now twice as big as before.

"Thereby the general theory of relativity is finally completed as a logical edifice," Einstein wrote on November 25, 1915. "The relativity postulate in its most general formulation, according to which space-time coordinates are physically insignificant parameters, leads with absolute necessity to a uniquely specified theory of gravity."[393] He had achieved a formula that could be condensed into a single line, with which he and other researchers could soon describe the entire universe. "It is the most valuable discovery of my entire life," he wrote to the physicist Arnold Sommerfeld.[394]

The fact that after his eight-year odyssey Einstein spoke about an "absolute necessity" is not without a certain comic element. He relativized this statement later: in light of the previously acquired knowledge, that which had been fortunately reached seemed almost obvious. "But only those who have experienced that themselves know the anxious, year-long search in the darkness, the alternating moods between confidence and fatigue, and the final breakthrough of truth."[395]

A Couple of Real Fellows

Back to Göttingen. Five days before Einstein's triumph, Hilbert submitted an even more extensive, far-reaching theory for publication. It encompassed gravitation as well as electrodynamics. The axiomatic structure of his paper was typical for the Göttingen School, which he had cultivated to a true mastery in this field. His mathematical method enabled him to reach a straight path to the field equations. But was Einstein's general theory of relativity already "ripe" for such an axiomatic approach before its completion?

This however did not apply to Mie's* theory of matter. If only for this reason, Hilbert's attempt to unite gravity with electrodynamics failed, even if his idea was revolutionary. Einstein himself followed Hilbert's footsteps a few years later—just as fruitlessly—trying to find a unifying field theory.

In regard to gravity, Hilbert was more successful. The conclud-

* In 1910 German physicist Gustav Mie attempted to derive electromagnetism, gravitation, and aspects of the emerging quantum theory from a single variational principle.

ing mathematical procedure he had applied, without which modern theoretical physics can hardly be imagined, had taken him very close to Einstein's solution. Until well into the 1990s, historians of science hardly doubted that his manuscript had already entailed the correct equations of gravitation. But in 1997, Hilbert's proof sheets from December 6, 1915, surfaced in the Göttingen University archives. What a spectacular finding!

Hilbert revised his manuscript in several places one more time after November 20. Among other changes, he backed away from an assumption that was in accordance with Einstein's earlier paper, "The Formal Foundation of the General Theory of Relativity"—not, however, with its final theory. The big surprise was that the field equations were missing. Hilbert made calculations up to a certain formula that is within reach of the gravity equations. But the actual equations were not there.

Or maybe they were? Opinions about this differ up to this day. There is a missing section from the, perhaps, decisive place in the proof sheets. Part of that page proof was cut off. What was on that scrap?

The fierce controversy that flared up shortly afterward displayed similar characteristics with the famous priority argument between Isaac Newton and Gottfried Wilhelm Leibniz about the discovery of differential calculus. Likewise, back then it was not the researchers but their admirers who picked up a quarrel. Hence Newton and Leibniz were dragged, after their pioneering achievement, into decades of sordid conflict, in which they participated quite ingloriously.[396]

Einstein and Hilbert were spared such a fate. Einstein seemed somewhat piqued only in the first moments. On November 26, 1915, one day after his triumph, he wrote to Heinrich Zangger that his new theory of gravity was of unrivalled beauty. Only one colleague truly understood it. "And that one tried to 'nostrify' it in a subtle manner."[397]

These lines appear in a letter to a friend in Zurich who was not part of the scientific community. The researchers themselves again contacted one another rapidly. Hilbert referred correctly in his publication to Einstein's academic works. Apparently, he was more concerned about his unifying approach and concluding mathematical procedure than about the announced field equations. Irrespective of what was written in the missing paragraph, he too considered Einstein the exclusive author of the general theory of relativity.

In December 1915, Hilbert recommended his Berlin colleague for membership in the Academy of Sciences in Göttingen. Einstein expressed his appreciation and alluded to "a certain strained mood" between them. He fought successfully against any acrimonious feelings. "I think of you again with unclouded friendship and ask you to attempt the same with me." It would be a shame if two "real fellows," who had managed to somewhat extricate themselves from this shabby world, could not find joy in each other's company.[398]

Big gestures, big minds, who maintained an open interaction with each other, full of humor. On his next visit to Göttingen, Einstein stayed with Hilbert, whose wife had been looking forward with anticipation to having the visitor. She was completely taken by Einstein's undemanding nature but did her best not to put it too much to the test, as her spouse confided to the visitor from Berlin.[399]

The differences between Hilbert and Einstein remained despite their mutual appreciation. They willingly rubbed against it. While Hilbert clung to his axiomatic approach and repeatedly reworked the unification of the general theory of relativity with Mie's electrodynamics, Einstein could not warm up to either of them. In one of his numerous academic papers he dealt with Hilbert's approach extensively.[400] In another place he called his approach childish: "just like a child, who is unaware of the pitfalls of the outside world."[401]

Einstein searched Hilbert's axiomatics in vain for a physical reference point. Furthermore, he was annoyed with the rather bold manner with which the mathematician had taken his theory of gravity and fused it with the structure of matter. Einstein considered physics too challenging to be left in the hands of mathematicians.

8. Shaking Spacetime

No Room for Black Holes

Following the excitement of those weeks in the fall of 1915, Einstein could not find rest. He immediately launched into further intellectual exploration, as though the discovery of the field equations of gravity placed him in a new continent of thought; his findings entailed new questions. Apart from that, he could delight how, in the midst of the war, the theory of general relativity awakened the interest of leading peers both in Germany and abroad.

He even received mail from the Russian front: Karl Schwarzschild grappled straight away with Einstein's publications "despite heavy cannon fire." The astrophysicist from Potsdam had mastered the big hurdles of non-Euclidean geometry fifteen years earlier. At that time, at a convention of the Astronomical Society in Heidelberg, he established the thesis that the cosmos is not flat but curved. In contrast to Einstein, however, he thought about a curved three-dimensional space and not a curved four-dimensional spacetime.

Schwarzschild was entirely fascinated by the mathematical consistency of the general theory of relativity, the almost unmanageable number of puzzle parts that join together into a set of mathematical formulas. In order to find a solution for the field equations, he first reduced it to a simple physical case: what does the gravitational field of a spherical star, similar to the sun and isolated in space, look like?

Schwarzschild assumed that the star did not rotate and that it was similar to an incompressible fluid ball. In this ideal case, he could actually calculate the external and internal gravitational field. His result, which up to this day is presented in every seminar about the theory of relativity as the "Schwarzschild Geometry," corresponded with the spacetime structure, in which a planet like Mercury follows an orbit that coincides with Einstein's already established approximation. A satisfying result for the astrophysicist. It was wonderful "that such an abstract theory should result in the explanation of the Mercury anomaly."[402]

Einstein congratulated his colleague for the first conclusive solution of the field equations and presented it to the academy on his behalf in the beginning of 1916. In the meantime, Schwarzschild submitted his calculations for the gravitational field inside the star. There was just one catch: it was impossible to solve the equations inside a specific radius. Could Einstein tell him anything about the physical interpretation of this barrier?

While Schwarzschild was waiting for an answer, he left the front due to pemphigus, a blistering autoimmune disease. His health deteriorated dramatically despite medical treatment. On May 19, 1916, Einstein's first official duty as the new chair of the German Physical Society was to announce the death of the astrophysicist at the age of forty-two.[403]

Since neither Einstein nor any other physicists knew how to start anything with Schwarzschild's formula, a good twenty years passed before the first researchers began to ascribe physical meaning to the "Schwarzschild radius": if a celestial body were to be condensed in this radius under the impact of gravity, then, for example, the entire mass of the sun would be concentrated within an inner ball of six kilometers in diameter, in which case the space curvature would be so big that not a single light ray could find its way outside. Any occurrences inside this horizon would remain hidden from an external observer, as though there were a membrane in one spot in the universe, through which light and material objects could invade but not escape.

Physicists today commonly talk about "black holes" and even divide them into different classes. The lightweight among them are "stellar" black holes, which have in them a maximum of a couple dozen solar masses. Astronomers discovered the first contender for such a black hole in the northern sky's Swan constellation. There, more than six thousand light years from Earth, a giant blue star whirls around a compact, invisible mass. In the process, the giant star continually loses matter, which then spirals toward the black hole and, as its heat increases enormously, emits conspicuous roentgen radiation. The object designated as Cygnus X1 attracted the attention of researchers in the 1960s as a strong source of X-rays. After lengthy observation with different instruments, they discovered that the giant star rotates around an invisible object of approximately fifteen solar masses.

Black holes of this sort form at a later stage of development from high-mass stars that have drained their nuclear fuel. Furthermore, astronomers have verified black holes in the center of the Milky Way and other galaxies. Millions, perhaps even billions of solar masses condense here in the narrowest space, what can be understood as basically a gradual growth in the course of the melting of galaxies and their cores. In the early universe, massive black holes could have been the hotbed of galaxy formation.

In 1916, however, such reflections were a long way away. At the beginning of modern cosmology and long before the emergence of nuclear physics, on which our present understanding of the formation of stars is based, only mathematics alluded to such exotic states of matter.

The complexity of the field equations made it difficult even for the creator of the general theory of relativity to distinguish between the physically possible and the impossible. In the course of time he found images for visualizing certain processes. "I compare space with a cloth floating (motionless) in the air," he wrote to his Dutch colleague Willem de Sitter.[404] If one pictures celestial bodies like balls rolling over this stretched cloth, then they dent the elastic cloth in their vicinity; in other words, they bend the space. The running dents affect the orbits of neighboring bodies as well.

What would happen if one were to suddenly place a celestial body in the universe? Einstein's simple cloth model suggests that the effect of gravity would not be immediately noticeable in the entire universe. The shock to the spacetime construction caused by the mass could spread on the cloth only with a finite speed in the form of a gravitational wave—with the speed of light, according to Einstein. After he had initially excluded the existence of such waves in his correspondence with Schwarzschild, a few months later Einstein was certain that gravitational waves were a necessary constituent of his theory. He understood them as a counterpart of electromagnetic waves.

Electromagnetic waves are emitted by accelerated electrical charge. At that time, such waves could already be broadcast over continents with the help of antennas. For instance, the big antenna in Nauen near Berlin established radio contact with Sayville in the USA. In 1916, it could transmit for express service 250 letters per minute over the ocean.

Einstein came to the conclusion that in places in the universe where masses were accelerated, energy would be likewise free, namely in the form of gravitational waves. In the geometry of space and time, such wandering distortions should propagate like electromagnetic waves with the speed of light and cause the distances between the bodies to periodically strain and compress. If, for example, a gravitational wave happened to come by while you were reading this book, your distance to the book's pages would change slightly.

Matter everywhere in the universe is accelerated; in other words, wherever and whenever something happens, gravitational waves emerge. The entire cosmos is presumably permeated by such waves.

Compared with electromagnetism, the impact of gravity is extremely weak. While electromagnetic waves generate themselves artificially and can be used for broadcasting information, this is not possible with gravitational waves. It would be entirely futile to want to receive gravitational waves emitted from any device.

Like all celestial bodies that orbit around one another, the earth and the sun incessantly emit gravitational waves, thankfully with just the minor power of 200 watts; hence, since the birth of the planetary system some 4.5 billion years ago, Earth has hardly lost any kinetic energy. Only over a period of calculated sextillion* years will this energy loss make itself noticeable, until Earth finally plunges into the sun.

Planets like our earth are cosmic lightweights. The effect would be greater when two stars rotate around each other. The higher the mass of the celestial bodies, the more energy is lost through gravitational waves.

Since the mid-seventies, astronomers have been observing double stars that emit gravitational waves. Back then, Russel Hulse and Joseph Taylor discovered a binary system of very compact neutron stars that rotate about each other in an orbital period of around eight hours. The two American scientists were able to precisely determine the orbital period using a radio telescope in Arecibo, Puerto Rico. Over the years they ascertained that the cycles did not remain the same but grew shorter.

The only convincing explanation for this is that the binary star system radiates gravitational waves and thereby continually loses energy, so that the two stars move closer together and faster around each other. All observation results support this thesis. The measured orbital decay was an exact match to the values ascertained by the general theory of relativity. The huge distance between the two neutron stars shrank by around 3.5 meters per year. In around 200 million years the two will merge and form a black hole.

Hulse and Taylor were honored with the Noble Prize for physics for their confirmation of Einstein's theory. Indeed, this is a matter of proving the gravitational waves indirectly. Could the mysterious spacetime vibrations be made visible in some way?

Say it came to a stellar explosion in our direct cosmic neighborhood. Then the produced gravitational waves would elicit tiny periodical length variations as they passed through our solar system.

* Earth will not reach this age; it will be engulfed by the sun in about five or six billion years, when the sun becomes a red giant.

Model calculations show that the huge distance between Earth and the sun would then vary by a miniscule amount, about the diameter of a hydrogen atom.

This did not deter physicists. For many decades, groups of researchers from around the world have been devising new means and ways to determine the tiny ripples of spacetime that come from supernova explosions in our galaxy or from other cosmic tremors. There are now huge gravitational wave antennas in different places on earth. When the passing of a gravitational wave causes a slight change in the length of the antenna arm, a sensitive laser device registers the tiny difference.

Physicists have also come up with a masterpiece of metrological detection. Precisely one hundred years after Einstein established his theory, a gravitational wave left a visible trace in a detector for the first time.

On September 15, 2015, Marco Drago stared incredulously at his monitor at the Max Planck Institute for Gravitational Physics in Hannover after he uploaded the newest readings to his computer. The scientist from Padua, Italy, could not trust his eyes. If the signal he saw in front of him was no artifact but had a physical background, then it was the most formidable explosion that astronomers had ever observed. He recalled how he couldn't believe his eyes.

The colleagues he consulted in Hannover and the United States were also initially skeptical. Their measurement devices were still at the testing phase. But the low frequency corresponded exactly to what the theoreticians expected when two black holes melt into each other: a short "chirp," comparable to a bird's.

Two laser systems with arms approximately four kilometers long, one in Hanford, Washington, and the other in Livingston, Louisiana, received the gravitational waves independently of one another. The signal arrived in Hanford seven milliseconds later than in Livingston, which is explained by the distance of 3,000 kilometers between the two antennas. An epochal discovery, even when no one still doubted the existence of gravitational waves.

The researchers made public their results only months later. The detailed data analysis led them to the conclusion that they had observed the merger of two black holes, one weighing 29 solar mass, the other 36. After a short circle dance they became one single black hole of fused 62 solar mass. And the other 3 solar masses? Within fractions of a second, in compliance with Einstein's famous formula $E=mc^2$, they transformed into pure energy. An incredible energy release! Momentarily, the firework surpassed the radiant flux of all the stars in the visible universe.

On Christmas 2015 researchers registered with the same antennas the merger of two additional black holes. The persistent search for Einstein's predicted gravitational waves was worthwhile. The measurement devices researchers have painstakingly built over decades now enable them to study cosmic processes invisible to conventional optical telescopes. The gravitational waves, which enable a unique insight into the early phase of the universe, must also have been originated by the Big Bang.

Not Without His Violin

Einstein's general theory of relativity has once again opened a new window to the universe. His physics colleague Max Born regarded the establishment of the theory as the "greatest achievement of human thought on nature, the most astonishing union of philosophical depth, physical intuition, and mathematical art." To the annoyance of Born's wife, Hedwig, he even took Einstein's work on gravity with him on his honeymoon. He considered it "like a work of art one can delight in and admire—from a proper distance."[405]

Hedwig Born also took the founder of the theory of relativity to her heart. From 1916 onward, Einstein regularly appeared on her doorstep with his violin in order to play music with her husband. According to the hostess, after entering the apartment, Einstein removed his "little rolls" (sleeve garters)—the loose cuffs of the thrifty man—and threw them in the corner. Then they played Haydn. "He radiated with goodwill." Einstein helped her to stop feeling as though she was lost on an icy moonscape when she was among natural scientists.[406]

Discharged from military service due to asthma, Max Born was an exceptional professor who came to Berlin in order to relieve Planck of his teaching duties. His first university semester at the Friedrich Wilhelm University had been halted prematurely for lack of students; consequently, he volunteered with the army. Now he researched the sound ranging of the enemy's artillery at the artillery assessment commission, which was located quite close to Einstein's apartment.

Thus began a period during which Born saw his colleague "very often, sometimes daily" and was allowed to observe the workings of his mind.[407] At the beginning of 1916 he wrote an article in a periodical on the general theory of relativity, and later an entire textbook about it, but he made the decision never to work in this field. Given the complexity of the field equations, he chose to observe the theory "from a proper distance."

The same applied for Planck as well, who had warned Einstein emphatically a few years earlier that the attempt to base the proven laws of gravity on a completely new conceptual foundation was a hopeless undertaking. However, just a few days after completing his general theory of relativity, Einstein was delighted to note that Planck was one of the first to take up the matter more seriously. "He still somewhat struggled with it. But he is a splendid man."[408]

In addition to belonging to the very rare species of theoretical physicists, Born, Planck, and Einstein were connected through a great love of music. Planck, renowned as an excellent piano player, regularly invited friends and acquaintances to house concerts in his villa in Grunewald. Einstein liked this form of socializing. He encountered in Mozart and Bach's compositions a harmony similar to that of the works of Maxwell and Newton. He walked through the city with violin case in hand and made music with Born and Planck, and sometimes also with Nernst or Freundlich.

Haber neither played any musical instrument nor had any desire for such gatherings during the war. Between his deployment at the front, the race for chemical warfare, and the manufacturing of explosives, he put himself under constant pressure. He was traveling unceasingly, trying to live three lives in one. Even when he stayed in Berlin, he managed many tasks simultaneously. At best after work he played a friendly round of chess—a competitive game that Einstein did not like—and read another thriller or serious literature, in order to fall asleep.[409]

But whenever he was able, Haber could be seen at the German Society 1914, a prominent club in the capital, where the industrialists Walther Rathenau and Robert Bosch, and the writers Thomas Mann and Gerhart Hauptmann socialized. But Haber was drawn to the Venetian-style Pringsheim Palace for a different reason yet, for that is where he met Charlotte Nathan, the secretary of the society, whom he married in the fall of 1917.

Compared to Haber, Einstein lived on an island of calm. He alternated between phases of intensive spiritual exertion and leisure hours, long pauses and holiday travels. "If many people pride themselves on having no time, Einstein was proud to always have time," noted his physics colleague Philipp Frank, an impression that is amplified by reading Einstein's letters and descriptions of those who met with him during the war years.[410]

Frank recounts meeting Einstein on the Potsdam Bridge in order to continue from there to visit the astrophysical observatory. "Since I was rather a stranger in Berlin, I could not promise to be there at an exact point." Einstein's response was that he would therefore wait on the bridge. Frank feared it would cost him a great deal of time. "Oh no," Einstein answered. "I can carry out the work I am doing at any random place. Why should I be less able to contemplate my problems on the Potsdam Bridge than at home?"[411]

Whether Einstein was standing on the bridge, walking in Grunewald, or sailing on the Wannsee, these say little about his real whereabouts. When he wished to dedicate himself to the pleasant

Image 8: *Entirely in his element: Einstein as lecturer.*

task of thinking, he was not bound to a designated place or time. As soon as he moved in the gravitational field of his thoughts, the categories of space and time lost their conventional meaning. "In real thinking, thoughts are closer to their fellow thoughts than the thinker is to his social environment," notes the philosopher Peter Sloterdijk. Therefore it is impossible to determine the place of thinking by providing ordinary topology.[412]

Einstein said about himself that he loved thinking as an aim in itself, just like music. The driving force of scientific thinking was not an external aim one aspired to, but the joy of thinking. "When I have no problem to think about, I like deducing again mathematical and physical propositions I am already familiar with," he wrote to his friend Heinrich Zangger.[413] Just in order to keep on thinking.

When Einstein said he was doing exceptionally well, it meant: "I live entirely secluded, working and silent."[414] Such being the case, it is understandable that it went against his grain that his mother and the rest of his relatives kept pressing him to marry Elsa, whose reputation, according to them, was at stake, as was that of her two daughters. He had already experienced one failed marriage, which in his view was a form of slavery dressed in cultural garb; why then should he risk the freedom he had grown to love so much?

Nevertheless he yielded. Out of a sense of duty for the daughters, he decided upon the "formality of marriage" with his cousin. Elsa had to accept that his life would not change in the least.[415]

In February Albert wrote to Mileva, telling her that he now wanted a divorce. He had no idea what his letter would trigger: his wife had a nervous breakdown for which he showed no understanding. Furious letters went back and forth between Berlin and Zurich in the ensuing months, alienating him further from his children. But Einstein's creative phase was unaffected. He remained within the flux of his thoughts, even in this tense situation.

How little importance Einstein attached to external circumstances is demonstrated by many daily incidents. For example, the student Rudolf Jakob Humm visited his apartment in Berlin-Wilmersdorf during the war and subsequently recorded in his diary: "Was this morning at Einstein's. He wore socks and put on sandals while speaking." First, Einstein finished reading a letter, then called a professor, and finally looked for the three physics papers Humm had asked for. Einstein went with him from one room to the other through a rather bare apartment, initially not quite wanting to come out of himself. But then he did not stop talking. The student admired the clarity and the all pervasiveness of his thoughts. Einstein almost never doubted; when he did, it was a clear doubt. "I stayed one and a half hours, outrageously long, and then I also accompanied him to a tobacconist's and from there up to the door of an acquaintance."[416]

Einstein talked about his physics with great serenity, initiating clueless students and neighbors, journalists and members of the Literary Society into the foundations of cosmology. His preference was to sojourn with fellow thinkers. Just as playing music, philosophizing with Planck and Born brought him consolation for the misery of the war and allowed him to occasionally forget their conflicting political views.

Einstein gave credit to his mentor Planck for having a tempering effect on other scholars and for not stopping the exchange of ideas with scientists abroad. It was thanks in no small part to Planck's persuasiveness that the Prussian Academy had not precluded its members from "hostile" countries. In the meantime, Planck had distanced himself so far from the appeal "To the Civilized World" that he was ready to submit an explanation meant for the Dutch press. In correspondence with the physicist Hendrik Antoon Lorentz and consulting with a few colleagues, he struggled for weeks to find the right words. His honest effort shows just how difficult it was even for moderate, patriotic scholars to take even one step toward the other side.

Annexationists versus Moderates

From the moment the talk of war flared up, the German professorate divided into two camps: on one side the strong faction of the annexationists, who constantly distributed new appeals in the universities and whose expansion wishes grew bigger instead of smaller during the course of the war; on the other side the "moderates," Planck among them. They stood just as well on "firm national ground" but rejected aggressive territorial claims—at least those concerning western Europe.[417] As in the summer of 1915 it came to an alliance of convenience between the moderates and the marginal group of pacifists; Planck and Einstein voted together against an illustrious circle of reactionary professors.

Planck continued to declare to the Academy that the German nation, which "has never believed in the Kaiser more unanimously, more sincerely, more joyfully," was forced into war at this fateful hour by envious enemies.[418] Why didn't he acknowledge the responsibility of the Germans for the war? Unlike Einstein, Planck did not challenge either the Prussian power and authoritarian state or the accompanying conservative national values. At most, he was ready to accept cautious reforms from above. The democratic renewal of Germany advocated by Einstein and other colleagues of the New Fatherland League went too far for him.

In February 1916, the League was banned; a month later, its director, the journalist Lilli Jannasch, was taken into "protective custody." Two weeks later she was released under the condition that she refrain from political activity during the war.[419] Other members of the organization had already gone through house searches and bans on travel.

Nothing of the sort is known about Einstein. But by now the military agencies had been keeping an eye on him for quite some time. A police investigation report about his pacifist activities was requested in December 1915 and submitted in January 1916. It stated that he had not attracted any attention for "fomenting so far."[420] In March, the Berlin commandant headquarter complained to the Prussian Academy that Einstein traveled around Germany without reporting to the police.[421]

The dissolution of the organized opposition, the censorship of the press, and increasing newspaper bans did not, however, silence the belligerent voices in the Reich but led to a radicalization of political conflicts. For example, on March 25, a few days after he caused tumult in the Reichstag with his opposition to further war credits and the warning that the war would leave behind no victors, only losers, the SPD politician Hugo Haase was forced to resign as chair of the party. Later, Haase, along with some other opponents of war, established a more left-leaning party, the USPD. A few weeks after that, on May 1, police arrested the anti-militaristic Karl Liebknecht. When the trial against the politician started early in the summer, over fifty thousand workers took to the streets, a prelude to politically motivated general strikes in the Reich's capital.

Right from the onset of the war, Einstein criticized the political conditions, above all from a moral standpoint. He did not join any party. On account of his Swiss citizenship he had already been keeping his distance from high politics.[422] Nonetheless, he made no secret of his pacifist convictions, even after the New Fatherland League was banned. For instance, in an obituary for Ernst Mach, published in April 1916 in the *Physikalische Zeitschrift* (Physical Journal), Einstein praised the philosopher for his friendly disposition toward people and his advocacy of understanding between nations. "This disposition protected him also from the disease of the time, from which only a few are spared today, namely national bigotry," wrote Einstein.[423]

One month later, he intervened in science policy for the first time when he succeeded Planck as chair of the German Physical Society. That same year, with "the highest writ" from Wilhelm II, Einstein was convened to the Physikalisch-Technische Reichsanstalt (National Metrology Institute), to which Planck and Nernst also belonged. Einstein wrote to his friend Zangger: "It is nice here and I swim at the 'top,' but alone, like a drop of oil on water, isolated by my disposition and view of life."[424] Hence he was inexpressibly delighted with each journey to Switzerland or Holland, although applying to go involved a great deal of effort.

Einstein's stance on the war was described in the scientific circles in Berlin as "naive." After a visit to Planck's house, the physicist Lise Meitner wrote: "Two superb trios (Schubert and Beethoven) were played. Einstein played the violin and shared in passing his deliciously naive and peculiar political and military views. The fact that there is an educated person who does not pick up a newspaper at all in these times is definitely a novelty."[425]

Apart from the fact that Einstein presumably subscribed to the liberal *Berliner Tageblatt*—as can be gathered at least from his police files[426]—the physicist was right that he paid relatively little attention to the censured press. Einstein said that often a fleeting glance at a newspaper was all it took in order for him to be "put off by his fellow man."[427] He preferred to familiarize himself with the international perspective by reading forbidden works such as the anti-war book *J'accuse* by Richard Grelling on the German share of the war guilt, which Einstein recommended to others; *Der Untertan* (*Man of Straw*) by Heinrich Mann; or the works of the Prussian court historian and antisemite Heinrich von Treitschke.[428] By the end of the nineteenth century, the Berliner university professor Treitschke had sparked an open debate about the position of Jews in German society and stigmatized them as opponents of national unity and as Germany's "misfortune." "This is the name one could use to label the local gentry," Einstein observed.[429]

His "deliciously naive and peculiar" political views resulted from his conviction that the fate of the civilized world depends on the moral forces "it is able to raise."[430] In any case, "his despair of the war and his extreme, fundamental pacifist views" provided topics for conversation.[431] Next to scientists and engineers, politicians and philosophers, artists and journalists were among the growing circle of acquaintances of the physicist, who was surrounded by the aura of a genius. He himself registered with some wonder how rapidly news of his revolutionary worldview spread in Berlin beyond the boundaries of his field. Thanks to his reputation, his fellow human beings regarded everything about him as good and fine, he ascertained in the summer of 1916. "This reputation bears sound endurance tests. But woe when it goes off."[432]

Walther Rathenau also became curious about Einstein's theory of relativity. As former director of the Materials Supply Division, Rathenau tightly organized the wartime economy in order to enforce German supremacy in Europe. Einstein confronted the industrial magnate and politician with his own democratic republican ideals. In a short note accepting Rathenau's dinner invitation, he rather unconventionally objected to Rathenau's claim that smaller states have no existence, countering that states bigger than the small region of Brandenburg have no right to exist. Einstein was obviously thinking of small Switzerland, specifically its sovereign cantons, as a model. He felt the state was necessary only as a carrier of nonprofit institutions such as hospitals, universities, and the police.[433]

The "Hell of Verdun"

The German Reich maintained field hospitals, military training camps, and armies. It waged a war that in 1916 caused more casualties than ever before in European history. The Supreme Army Command prepared the disastrous plan to "bleed French troops white," to attack them with heavy artillery at their military outposts, which they could not possibly surrender, until their will to fight was broken. The failed strategy entered into world history as the "Hell of Verdun." Within the first three weeks of combat, the German artillery fired some three million shells in this place of horror, an unbelievable compression of violence within the narrowest space.[434] According to the writer Arthur Holitscher from the New Fatherland League, thirty-seven railroad trains with fourteen wagons full of ammunition drove up to take the bulge of Verdun, a strategically important elevation of 304 meters. "These details were supposed to save the people from defeatism, to prove the seriousness and efficiency of the nation."[435]

For France, the battle was a fight for survival, and the commander Philippe Pétain was "the nation's savior." He opted for a defensive strategy based on the continuous rotation of troops. Consequently, almost 80 percent of all French regiments took part in the battle of Verdun, which ultimately strengthened enormously the will to wage a defensive war.[436] Admittedly the Germans did temporarily seize the big strongholds of Douaumont and Vaux. But due to a well-organized and ferocious resistance, the attacks on other strongholds led to enormous losses on both sides. In Verdun, 320,000 French soldiers and 280,000 German soldiers were killed, injured, taken prisoner, or went missing. Planck's older son was killed there as well.

An even bloodier war of attrition began that same summer at Somme, where the Allies' forces attacked German positions with ceaseless artillery fire. Here too, the offensive did not result in the collapse of the front. With its fortified positions and machine guns, the defense once more had an advantage over the advancing infantry.

On July 1 alone, the British grieved twenty thousand dead and twice as many wounded.

Max Born left in the summer of 1916 to inspect stations for artillery sound ranging measurements at Somme, where he too got into battery fire. Unlike his comrades, the asthmatic could not bear the air in the dugout for long. He took position on a mound, from which he could oversee the fighting over many kilometers, "a terrible wall of smoke and fire combined with an infernal noise," under which the acoustic measurement method collapsed. Born observed several "stirring aerial combats" and had the feeling of attending a play he had to record with all his senses. Only later did it become clear to him that he "had experienced the dreadful mass murder the war has degenerated into."[437]

Even in the midst of the worst phase of the destruction, Einstein became absorbed in physics. He gave a lecture to the German Physical Society on flying and was positive he thereby made an important contribution to aerodynamics.[438] Much more important was his work on quantum theory, which he presented at the same place in the summer of 1916. First he elucidated the processes that later became fundamental for the development of laser technology.[439] After this excursion into the microcosm, he returned to the field equations of gravity, with which he wanted to describe the spacious structure of the universe. For months he intensively contemplated the structure of the cosmos and yet remained alert to the atrocities of the war and the suffering of the people.

There was hardly a German who had not lost someone from his or her immediate family during the war. Facing the extent of loses on the Western Front, a simultaneous Russian offensive, Rumania's entry into the war, and the ever more threatening food shortage, the German people's confidence in a victory was deeply shattered.[440] Before the war, the empire had been the world's biggest importer of wheat, forage, and other agrarian products. This changed abruptly in August 1914. The British embargo drastically reduced imports. Farmers also lacked fertilizer, cart horses, and workforce, whereby the shortage of provisions was further aggravated. Bitter poverty prevailed in Berlin's tenement houses.

As early as January 1915 one could only get bread in the capital with ration cards. To stretch out their supply of bread, bakers resorted to rutabagas, potatoes, corn, or sawdust. Butter, sugar, and meat were likewise rationed. Next to a Reich's legume center, there was also a Reich's distribution center for eggs and a Reich's commissioner for fish supply. But even the War Office of Food no longer succeeded in ensuring the minimum supply of essentials to the population. Hunger and food shortages defined everyday life in the big cities. The black market flourished but could only offer "disappointing substitute goods."[441]

The privileged, such as Walther Nernst, could keep on supporting themselves with game from their own estates, which the chemist had delivered to a Berlin cold storage where he rented space.[442] Max Born, by contrast, had already been through a nasty "rutabaga winter" in 1916. "Rutabagas served as everything, not just as vegetables, but also as substitute for marmalade, as admixture for flour in bread and cakes, and I do not know as what else."[443] His wife Hedwig was not able to get back on her feet for a long time after the birth of their daughter, due to malnutrition.

In the two final years of the war, during which he was constantly sick, Einstein frequently had to ask relatives and friends for aid packages. In the meantime, the lineups at the food stores gave him some hope of speedy armistice negotiations. "The war has a beneficial educational effect on the people, or better formulated, the food shortage," he noted in July 1916 to Zangger. "If it keeps on going like this systematically...the fellows will become more agreeable."[444]

The Total War

One month later, in August 1916, the Chief of the General Staff, Erich von Falkenhayn, was replaced. The new Supreme Army Command, motivated by the glorified "liberator of East Prussia" and "the Victor of Tannenberg," Paul von Hindenburg, and his Chief of General Staff, the military technocrat Erich Ludendorff, issued a new armament program two days after taking over the administration. They wanted the production of ammunition to double between now and the spring of 1917, a massive extension of the manufacturing of canons and machine guns.

On the Eastern Front, Hindenburg and Ludendorff had confronted a numerically superior but technically inferior Russian force. Now they wanted to offset the decreasing number of soldiers fit for combat with further technical means. They relied on the quality of German artillery, submarines, and chemical weapons.

The theologian and philosopher Ernst Troeltsch, with whom Einstein was in contact, considered this overestimation of technical requirements to be the essential reason for the prolongation of fighting. The war had turned into a "vicious spiral."[445] Meanwhile, Walther Nernst warned that one should not ask the impossible: "But it seems to me an impossible task, now in wartime, to accomplish so to speak overnight a submarine fleet with the appropriate resources to go with it, and of the required range, in order to radically, but truly radically, intervene in the course of the war."[446]

The discussion of a submarine warfare led to even deeper rifts in the academic world. Those endorsing an aggravated submarine war wanted to operate the new weapon system against British merchant ships and hit the cause of the embargo. The German navy had already started such an economic war in 1915 but stopped it after a few months. By firing at ships on which numerous American passengers were killed, they were risking bringing the USA into the war.

How was this risk to be assessed a year later? Was the USA prepared to intervene in the fighting on the European continent within months? What could German submarines accomplish against the British naval power? To what extent could Germany's own supply bottlenecks be conquered by a submarine war?

Nernst deemed it downright dangerous to foster illusions about these political and military questions without sufficient expert knowledge.[447] Particularly those rumors that circulated among the population about the alleged potential of the submarines. But once again, humanities scholars led a propaganda campaign that was ultimately aimed at the hesitant Reich's chancellor, Bethmann Hollweg. Historians and classic philologists, Egyptologists, and archivists believed that all the available weapons now had to be used with abandon. "We do not have to hold out, we have to win."[448]

In the fall of 1916 Einstein joined a new pacifist group, the Association of the Like-Minded, whose meetings were only partially documented. Its guiding principle: "vanquish nationalism, which contradicts the spirit of humanity." Ethical concerns instead of power politics should be taken into consideration in world politics.[449] Some of his comrades from the New Fatherland League belonged to this new organization. It worked quietly and wanted to influence public opinion with well-aimed actions.

One day, Max Born also received an invitation to the Villa in Tiergarten, where the members met regularly. There he met his colleague Einstein in a circle of about twenty intellectuals. "I was informed that the purpose of the gathering was to discuss the submarine problem with some high officials from the foreign office."[450] After an introductory lecture, a lively debate flared up. Although as an officer Born was obligated to keep away from secret political activities, he participated repeatedly in the meetings over the following weeks. "But our attempt to exert moderate influence led to nothing." In the end, Ludendorff carried through with his intentions.[451]

The third Supreme Army Command also expected speedy progress from German scientists in the field of chemical weapons. In June 1916, during just one night in Verdun, field howitzers and canons fired 116,000 shells filled with the choking agent diphosgene.[452] The massive bombardment spread fear and the terror of a slow, agonizing death. It also contributed to the determination of the French population to protect their country to the last man against such a brutal opponent.

With the intensification of chemical warfare at the end of 1916, Fritz Haber's institute was entirely in the hands of the military leadership. From this time onward, Berlin-Dahlem swarmed "with an odd amount of centaur-like beings," half officers, half chemists.[453] The number of scientific employees and assistants increased to more than 1,500. They had to be lodged in barracks built for that purpose, as well as in the Kaiser Wilhelm Institute.

The palette of chemical warfare being researched was likewise increased. For example, from 1916 onward Berlin researchers alone tested hundreds of different chemical bonds of the best known poison: arsenic. One result of their studies, titled "mask breaker," was a nose and throat combat agent that was supposed to force soldiers to pull off their gas masks. Those affected would then be completely exposed to other gases in a so-called "colorful firing," a term referring to the colorful marking of the shells filled with different combat agents.

"The constant work on strong poisonous substances had deadened us to such an extent that when we used it at the front we had no scruples at all," wrote Otto Hahn in his autobiography that appeared in the 1960s, in which—in contrast to other colleagues—he acknowledged his role in chemical warfare.[454] As an observer at the front he only seldom saw the effect of chemical weapons. But the sight of slowly dying soldiers made him feel deeply ashamed and made him aware of the senselessness of war. "First one tries to eliminate the unknown person in the enemy trenches, but when one faces him eye to eye one cannot bear the sight and helps him again."[455]

Did the sight of gassed soldiers stir similar emotions in Haber? There are no such indications, either in the correspondence with his second wife, Charlotte Nathan, or in his correspondence with researchers and industrialists. Instead, in his letters he chose different forms of rationalization and abstraction to keep at bay the terror of battle, including verse:

It's fine to serve the fatherland
Ammunition decorates the man
To serve the state is thrilling
And a good deed is fulfilling.[456]

To the millions of odes to war that had been composed in Germany since August 1914, Haber added a couple. His war poetry, largely ignored by his biographers, is worth mentioning because it slipped into the highest administrative level of the chemical warfare organization. Checking the archives sends chills down one's spine when one reads how Haber kept the industrialist Carl Duisberg in good spirits with poems from the War Office.

Nevertheless it would be a mistake to believe that Haber planned the gas warfare with a feather quill at his desk. The chemist meticulously got to the bottom of all the facets of deploying chemical weapons. He spent time in the laboratory, where, with the assistance of animal experiments, he determined the "lethal concentration" of different gases; at German enterprises, where he assessed the efficiency of industrial processes; and at the front, where the enemy's unexploded shells were analyzed directly onsite, in order to prepare German soldiers for coming into contact with new toxic substances.

The British and French also established big research programs for chemical warfare. But because of their outdated chemical industry, they remained inferior to the Germans in the race for chemical weapons and gas protection. Immediately after the attack at Ypres, fearing retaliatory measures, the German Chief of General Staff called for improvements to protective measures. Within the year German soldiers had at their disposal gas masks with exchangeable filters. While the allied armed forces donned long gauze bandages, helmets, balaclavas, and masks without filters—a completely insufficient cover against the always more toxic chemical warfare agents—Haber consistently developed further gas protection. He wrote the following verse to Duisberg:

In the army, still in use,
One breathes without a breathing tube.
The collar inset is still screwed
And a valve unapproved.
Hence the opportunity
One will seek immunity
Consider this a key protection
And pay the hose its due attention.[457]

Haber attached these words to an application for research work on the production of synthetic rubber. It was a matter of course for Duisberg "that requests presented with such wonderful rhymes are immediately granted." Haber's poems will "be preserved at the war museum of the paint factory in Leverkusen as great and valuable wartime memorabilia."[458]

As a result of the chemist's "mindfulness" and capacity for invention, the progressive terror evolved into an insane race for chemical warfare. In the meantime, the deployment of poison gas at the Italian and Russian fronts contributed considerably to Central Powers' victories. Among Russian soldiers alone there were some half million gas victims over the course of World War I.[459]

The use of mustard gas in 1917 proved especially devastating. Because the droplets stuck everywhere, a large-scale application of this poison could, under certain circumstances, cordon off entire sectors of the front. It rapidly attacked skin and eyes, causing painful blisters. Many of the poisoned—fourteen thousand British soldiers within the first three weeks—had to be led, blinded, from the battlefield.[460]

Haber's own employees suffered poisoning during tests with gas masks or when filling poison gas shells; mustard gas was introduced on the German side without sufficient provision of protective measures. Many of the more than two thousand workers, male and female, who toward the end of the war filled twenty thousand shells a day with mustard gas in Berlin-Adlershoff, fell seriously ill.

Magazines were now increasingly replacing human faces with gas

Image 9: *Gas masks for mule and rider*

masks. The gas mask became a primary symbol of the war. But in contrast to the unrestricted submarine warfare, public debates about chemical warfare were nonexistent. Neither do we know Einstein's views on this matter.

9. Einstein's Universe

His "Biggest Blunder"

"The Lord no longer needs to rain down brimstone and fire," Einstein wrote on February 4, 1917, to Paul Ehrenfest. "He got himself modernized and made this business run automatically."[461] The latent sarcasm in his correspondence can be seen as an indication that his thinking about everyday horrors was coming to an end. His comments on war-related events are confined in most of the letters to two or three lines. It is only in their entirety that they reveal the great impact the destructive power of the war had on his emotional state.

Most of his correspondence was reserved for research. In the above quoted letter to Ehrenfest, Einstein announced his next scientific publication: "I have once again perpetrated something in the theory of gravity, which exposes me a bit to the danger of being committed to a madhouse."[462]

Four days later, on February 8, 1917, he read to the Prussian Academy what he had perpetrated. Precisely one week after the declaration of unrestricted submarine warfare, Einstein submitted to the Academy his "Cosmological Observations of the General Theory of Relativity."[463] He thereby initiated a debate that continues to this day. Researchers are still trying to puzzle out the meaning of the "cosmological constant" that Einstein brought into the world on that day. Is this subsequent extension of the field equations a stroke of genius or his greatest blunder?

Einstein's image of cosmic structure is based on a simple thought: that the universe looks the same in all directions, and this must be the case for every observer in the universe. A relativistic point of view typical of Einstein. But this only partly matches the experience of a stargazer. When one watches the clear night sky away from the brightly lit cities, one can see single stars here, entire clusters of stars there—and furthermore, the band of the Milky Way. Doesn't it rather look as if all the stars form the Milky Way, some sort of concentrated galactic island beyond which there is nothing but empty space?

Einstein certainly considered that. In his view, Newton's theory of gravity requires the universe to have a center, where most of the stars bundle up, while stellar density decreases outward, creating an infinite empty space.[464] He himself was not pleased with the idea of an island composed of stars in the infinite ocean of space. For one thing, this would lend a meaning to empty space that contradicted his own theory. Besides, such an order would be impermanent. As Einstein explained, the light emitted from the stars as well as the stars themselves would little by little leave the islands and disappear into infinity without ever returning. One star after another would be projected out of the center in this gravitational game of celestial bodies, and indeed according to the laws of statistical mechanics, "until the entire energy of the star system is great enough—when transferred to a single celestial body—to allow it to travel into infinity, from where it will never be able to return."[465] The islands would be systematically impoverished.

According to Einstein there is no indication of such an evolution. For him, the leeway for cosmological models is limited, partly because the velocity of the stars in relation to one another is small. So small, that the stars do not spread and their constellations have been known for thousands of years.

At the beginning of the twentieth century, astronomers' horizon was more or less limited to the Milky Way. Researchers were not yet sure about the nature of cosmic nebula and knew nothing about other galaxies. To solve his field equations for the entire universe mathematically, Einstein had to look much further out than astronomers with their telescopes. He had to know if and how matter changed at an always a greater distance from Earth. Into infinite space.

To address this dilemma, he proceeded on a bumpy road. Einstein considered a few theoretical reflections for the hypothesis that the fixed stars are distributed evenly in space.[466] He thus assumed that "the density of matter is indeed very different in details, but on a large scale the average is the same everywhere."[467] It does not matter what direction one looks. "Expressed differently: as far as one may travel through space there are to be found everywhere loose swarms of fixed stars about the same kind and same density."[468]

According to the physicist Max Born "what was most striking about his way of thinking was his belief in the simplicity of fundamental laws."[469] This simplicity included all the possible observers, which is why he immediately expanded his reflections.

The point of view from which humans observe the universe is in no way privileged. Einstein gave a new dimension to Copernican thought: humans do not stand in the center of the universe, because such a middle does not even exist. A pivotal thought of modern cosmology.

From now onward his thoughts fit together with the help of fewer hypotheses into a new world model. First he made a far-reaching assumption about the geometrical structure of the universe. "The bending character of space varies temporally and locally according to the distribution of matter." But despite these local differences, geometry on a large scale can be "approximated through a spherical space." From the standpoint of general relativity this version seemed likely and was logically consistent.[470]

Einstein assumed that the cosmos is so massive that the existing matter bends space so starkly that it closes like a ball. Such a universe would be enclosed and yet infinite. An ant running on a ball but never coming up against a border can serve as an illustration. Since it always perceives only a little piece of the surface, it appears to the ant, locally, to be flat, although it is in fact curved.

Einstein's model universe was compatible with the astronomical knowledge of his time. Above all it allowed him to avoid the above-mentioned dilemma. If matter is actually distributed evenly over an enclosed universe, then there is no need for a special knowledge about the density of the stars far away in order to solve the field equations. Instead, it suffices to know how high the mean density of matter is.

Einstein, however, avoided another difficulty only with a stratagem. To his own dismay he had to admit that his relativistic field equations published in November 1915 did not allow for an unchanging universe; according to his own theory, the cosmos could not be stable. Einstein therefore decided to take a bold step: he once again modified the field equations on which he had worked for eight years.

We should pause here and consider Einstein's intellectual flexibility.

The general theory of relativity, for which he soon became world famous, was just one year old, and there he was putting it through the mill. His curiosity drove him not just forward from one thought to another, but also from previous theorems to their opposite standpoints. We have repeatedly seen in this book how he used internal discourse to search after what could not be appropriated into big theories. He turned this freedom of thought, a mocker of authority, against his own ideas.

From the standpoint of mathematics, the subsequent modification of the theory is plausible. Einstein inserted an additional part into the field equations: the "cosmological constant." This additional term has a deeper significance. "It shows that in 1915, in contrast to what he wrote, Einstein had not as yet found the most general equations that were compatible with his claims," explains the Einstein expert Jürgen Renn, director at the Max Planck Institute for the History of Science in Berlin. "The additional part plays a key role in modern science. It is indispensable for the explanation of the accelerating expansion of the universe."[471]

To be sure, at the beginning of the twentieth century Einstein knew nothing yet about the expansion of the universe, which is backed today by astronomical observations. By introducing the "cosmological constant" he followed the reverse strategy: he threw it onto the scale in order to achieve an equilibrated solution, namely a universe that neither expands nor collapses but remains stable.

At the beginning of March, while he was lying sick in bed for several weeks, he apologized to his Dutch colleague Willem de Sitter: "From astronomy's point of view, I have built nothing but a spacious castle in the air." As a theoretician he was mostly concerned about unspooling his theory of relativity to an end without running into inner contradictions.[472] De Sitter, however, doubted whether the cosmos model was actually as stable as Einstein believed.

With that small additional term, Einstein undermined not only his previous theory but all of physics. He channeled a sort of anti-gravity into the broad understanding of natural science. The "cosmological constant" corresponds to a field with a repulsion effect that Einstein could not link to any known physical phenomena. Even so he assumed that this tendency to repulse is somehow woven into the texture of the spacetime fabric.

Years later he described the "cosmological constant" as his "biggest blunder" and abandoned it. But he was wrong about rejecting it. In modern cosmology, Einstein's "biggest blunder" turned out to be one of his greatest predictions.

It became apparent that the universe expands and that this expansion is even accelerated—with the peculiar consequences that the remotest (though today still observable) galaxies will disappear from the horizon of future observers, should this accelerated expansion continue. It remains a mystery what "dark energy," what anti-gravity force, drives the expansion of the cosmos. It appears that the "dark energy" is distributed completely evenly over the entire universe. Besides, present observations indicate that "dark energy" has had the same density since the very beginning. Should it be confirmed that the energy density has remained constant for 13.8 billion years, then "dark energy" would be equated with Einstein's "cosmological constant."

Genius in Descent

As Einstein built his "spacious castle in the air," an era of aviation in Germany came to its end. Count Ferdinand von Zeppelin, who had captivated the entire country with his airships, died on March 8, 1917, at a hospital in Berlin. Less than ten years earlier, the crash of a 136-meter airship from the House of Zeppelin triggered the empire's biggest voluntary campaign for donations and laid the foundation for building an enterprise that until 1914 had transported around thirty-five thousand passengers. Except for a few military officers, who had thought the airships would soon be dropping bombs over London?

If at the onset of the war Zeppelins were regarded by many as German "wonder weapons," they had since lost that aura. They offered the heavily armed opponent much too easy a target and were difficult to navigate. Count Zeppelin eventually took part in the development of a huge airplane with a 40-meter wingspan that could carry two thousand kilograms of bombs.

Airplanes, air ships, and balloons were above all used for reconnaissance. They would soar over the front as the artillery's eyes and spy on the opposing positions. To prevent this, fighter pilots on both sides attacked reconnaissance aircrafts and each other, in what Max Born witnessed as the "stirring aerial battle" at the Somme.

In a war that no longer knew any close combat, and in which most soldiers were killed without ever having seen the "enemy," fighter pilots took the place of heroes. The army decorated them, and the press celebrated them as "Knights of the Air," using their names to entice Berliners into the big "German Aerial War Trophies Exhibition," which could be seen from February to April 1917 at the zoological garden. The exhibition organizer built trenches and wire entanglements in front of a mural, 450 square meters in size, that showed the pilots in action.[473] Every Berliner schoolboy knew the celebrated pilots: Oswald Boelcke, recently laid to rest in a state funeral, and Manfred von Richthofen, who had completed twenty-five air combats. With their agile machines, bi- and monoplanes, the fighter pilots performed flips "as if each of them were a Pégoud."[474]

The name of the Frenchman, who before the war had made headlines with his loops, was still on everyone's lips. But he too had died. At the age of twenty-six, in 1915, he was shot down near Petit-Croix by a German pilot, Corporal Walter Kandulski, who returned to the scene after the battle and laid a wreath on the crash site. He had cheered Pégoud before the war and had learned flying from him.

For Einstein, flying—overcoming gravity and soaring over earth—was still one of the biggest riddles of physics. His curiosity led him to try to find out what the load-bearing capacity of soaring machines was based on. "There is great uncertainty surrounding these questions," he prefaced an article in a popular scientific magazine in the summer of 1916. "Yes, I have to admit that I have not found its simplest answer, even in the specialist literature."[475]

As Einstein prepared to close this supposed gap, he could be certain of public attention. Soon after his lecture to the German Physical Society on fluid physics and aerodynamics and his corresponding magazine article, he came into contact with the LVG, one of the biggest aircraft manufacturers based in Berlin-Johannisthal. It is possible that a Berliner investor introduced him to the manager of the LVG at the time, Romeo Wankmüller.[476] In March 1917, the company tested an airfoil designed by Einstein. This "humped wing" would have fallen into oblivion, but thirty-seven years later, just a few months before Einstein's death, a German pilot and technician wrote to the inventor. "Honorable Professor," begins Paul Georg Ehrhardt's rather humorous letter. "With this letter I wish nothing else than to record what has been for me an unforgettable, for you perhaps forgotten, experience."[477] This experience, which Einstein could well recollect, went back to the time when Ehrhardt was sent to the LVG experimental department as technical manager in order to build a one-seater fighter plane. As a result he got the "ungrateful task" of having to deal with offers from inventors, a job Einstein knew all too well from the Bern patent office.

"I felt therefore somewhat disapproving when one day a piece of writing consisting of several pages found its way to my desk, moreover written by hand," Ehrhardt continued in his letter. "But as soon as I started to skim the impressively elaborate work, I realized that its writer had at his disposal a much higher knowledge of theoretical physics than I."[478]

It might have flattered Einstein how seriously his theoretical treatise had been taken at the time. The enterprise decided to test whether his thoughts would prove workable. Had Berlin's most significant physicist actually uncovered a secret of flying?

In March 1917 the LVG sent the wing profile manufactured according to his guidelines to Göttingen. The peculiar upward curved "humped wing," shaped like a cat's back and marked LVG D9, was tested in the wind tunnel at the experimental station for aerodynamics.[479] A few weeks later the company assembled a "cat's back" wing to the body of a LVG biplane.

"I presided over the task of putting it to the test in flying," continued Ehrhardt. At the time, each takeoff with a new prototype was a genuinely risky game. For that reason Ehrhardt had supervised the completion of the wing with growing skepticism. He feared that the machine would leave the ground with the most unstable flight attitude and would sag backward because the wing lacked the pitch angle. Indeed, this is what unfortunately transpired. "For after taking off I went airborne like a 'pregnant duck' and after an embarrassingly straight flight I was very glad to have the wheels on solid ground again, shortly before the airfield's end at the fence of Berlin-Adlershof."[480]

Ehrhardt was not the only test pilot to have risked going up with Einstein's "cat's back" wings. After the wing was rebuilt so that it had a pitch angle, the second pilot could dare to fly a curve. But he too was unable to pull off an orderly flight. "The pregnant duck was now a lame one."[481]

"That is what happens to someone who thinks a lot, but reads little," Einstein answered immediately on September 7, 1954. In his witty reply he described once again how he had arrived at the Bernoulli equation for the humped wing. Yet it did not follow from all this that it would be reasonable to form a wing in such a way. "Nature knew why it rounded the bird's wing at the front and made it sharp edged at the back." He had to admit that he was still embarrassed by his recklessness at the time.[482]

The LVG tolerated the failure. The whole affair ended with a small party, at which Einstein read from his theory of relativity. For the New Year the company sent the professor photos of the trials, for which he thanked them. Reviewing the photographic records, he would later find his excursion into the realm of practice not without humor.[483]

Just with humor? Did Einstein not know that that excursion of his during the war was actually an excursion into the military realm?

"A city teeming with life and with considerable dimensions" emerged in Berlin-Johannisthal during the years of the war. Its establishment looked a bit like the Wild West, wrote the journalist Arthur Fürst. Toward the end of the war, it occurred to the journalist how many famous scholars one met at the barracks, who collaborated quietly and humbly in the defense of the fatherland.[484]

Like all the companies located in Johannisthal, the LVG depended on armament contracts. Its war production went up from 600 airplanes in 1914 to 1,800 in 1918.[485] During this time the enterprise developed numerous military airplanes. The biplane, which in 1917 was provided with the Einsteinian airfoil, was an older model, but it was one of the first airplanes that dropped bombs on the English capital after the Zeppelins. How would Einstein have reacted if in the last years of war an improved LVG biplane with curved wings and a heavy bomb load had flown to London? Or Paris, where on January 30 and 31, 1918, sixty-five civilians were killed after more than 250 bombs were dropped?[486]

Did he not see or want to see the possible consequences of a collaboration with the aircraft industry in those days?

One would like to grant that he was above all interested in uncovering the secret of flying and finding out whether his calculations of it were correct. But the "pregnant duck" was not the only escapade of this kind. Earlier that same year Einstein had become a "scientific collaborator" with the Mercur Flugzeugbau GmbH (Mercur Aircraft Construction Ltd).[487] Its director, Romeo Wankmüller, who had left the LVG, incorporated Einstein's prestigious name in the firm's publicity. Furthermore, he hoped to receive some hints from the scientist for building a *Hebeluftshif*, an airship that used compressed air for takeoff.[488]

In any event, Einstein saw no clear boundary between theoretical and practical research. He was a pacifist—but for radical pacifists, Berlin in those years was the worst conceivable place, and Einstein did not wish to be one of them, otherwise he would have left the empire's capital a long time before. To avoid being caught in military matters during a time of total mobilization, he should have not only kept away from any industry, but from public life altogether.

Instead, he sat on the board of trustees of the Imperial Physical-Technical Institute, which conducted relevant military research. In October 1917, he undertook the management of the Kaiser Wilhelm Institute for Physics, which was newly launched despite the war and whose board of directors also included Haber and Nernst. Besides, he had already been dealing as an expert for some time with the gyrocompass, an instrument highly interesting to theoretical physicists and the former patent office clerk, but also a very important one for the navy. In the last years of war, Einstein prepared a private expert's report for Anschütz, the company that built such instruments.

Certainly, pacifism represented values that were very important to him, but they were not the only ones he followed. Einstein moved in different spheres with partly conflicting attributes. His decision to remain in Berlin went inevitably hand in hand with having to live with numerous contradictions and choose between different evils. He was not willing to substantially restrict his scientific curiosity, his own life interests, nor did he want to turn away from the scientific circles on whose account he had come to Berlin. Pacifism was a wing for his exploratory scientific flight, but a humped wing.

His freedom of thought did not live well with rigid value systems. How can we otherwise understand his friendship with Born, who, like countless other physicists, was directly involved in war research? His close relation to Planck who justified the war in his speeches? His connection to Nernst? His relationship with Haber, with whom he had maintained a close personal relationship before the war and through whose desk the Kaiser Wilhelm Institute research proposals for physics now went?

At the beginning of 1917, Haber made a couple of appearances at the Academy—clearly not for the sake of science, but in order to award Leopold Koppel, his most important industrial partner, the Leibniz Medal, for which Haber had submitted the application.[489] Afterward, Einstein and Planck, who attended weekly meetings at the Academy, did not see Haber again until the end of the war. In October 1917 Einstein congratulated him on his wedding. The congratulatory letter was not preserved, but Charlotte Haber informed her spouse, who just four days after the wedding ceremony had begun traveling again, that "Einstein wrote very nicely. He is pleased that I am a daughter of the Israelite people and no long-legged, aristocratic German."[490] It is doubtful whether Haber thought these lines were "nice." Charlotte Nathan, "daughter of the Israelite people," had converted to Protestantism solely upon his demand, which she regretted for the rest of her life. She had done so in order to be married in the Kaiser Wilhelm Memorial Church; as she said, Haber "would not be happy with less!"[491]

Prickly on the outside yet benevolent—that is how Einstein appeared to many. With all his acuity, he rarely passed judgment on anyone. As he wrote to his friend in Holland, after he found out that Nernst's second son, a pilot, was also killed, he had become more tolerant during the war, without basically changing his views in the least.[492] He did not wish to allow himself to be contaminated or influenced by hatred. He pursued science in a society entirely imbued by military spirit; he followed his scientific path not always entirely free of cynicism, but with much sobriety and without letting himself be influenced by others.

Einstein blamed the pervasiveness of military spirit, not his colleagues. He advocated structural reforms—for example, changing the educational system. In a notable contribution to the *Berliner Tageblatt* at the end of 1917, Einstein demanded the abolishment of all graduating exams, which, in his opinion, took all of the students' focus, thereby reducing the quality of classes.[493] He wrote to Romain Rolland that he resisted any "dressage" or soldierly education, which he felt misled people into believing that they existed for the state and that its power was an end in itself.[494]

In his view, apart from the educational system, the political structures had to change in order to put a stop to the destructive mechanism of highly armed nation states. And they would only change when enforced by the hardship of facts—a hope Einstein never gave up. "There is an epidemic-like delusion, which, after producing endless suffering, will disappear again, to be viewed by the next generation as something entirely monstrous and incredible."[495]

"November 9, 1918—Canceled due to revolution": Einstein the Activist

Missed Dress Rehearsal

"The negotiations with Trotsky begin again today, and 300,000 workers strike in Berlin," wrote the diplomat and diarist Harry Graf Kessler on Tuesday, January 29, 1918. "We are potentially standing before one of the most awful hours in German history; it might be nothing, but it could be the start of the German revolution."[496]

The strikers laid down their work in ammunition factories, mechanical engineering companies, and in the aircraft industry and carried signs saying "Peace! Freedom! Bread!" There were more protesters than ever before. And unlike in the previous year, it was not mainly hunger and necessity that drove them out into the streets. This time the workers were driven by political goals. Their protest was aimed against the war and the ruling regime.

The demonstrations in Berlin and other German cities were related to the revolution in Russia. The Russian Empire had collapsed. The Bolsheviks assumed power and called upon all the fighting nations to make peace. Its ambassador, Leo Trotsky, had been negotiating peace since December in Brest-Litovsk with the German-Austrian delegation that posed unacceptable demands. Trotsky prolonged the talks. He counted on the spread of the Russian revolution through middle Europe.

The general strikes in Germany seemed to prove him right. The Independent Social Democratic Party (USPD), which until then had orchestrated the biggest anti-war demonstrations, referred in its leaflets directly to the events in Russia: after claiming this was a defensive war, German militarism had finally unmasked its true intentions at the negotiations in Brest-Litovsk. "Robbing foreign countries...and putting the world under the rule of the German sabre: these are the war aims of the German government."[497] Against it, the leftist workforce demanded "a speedy initiation of peace without annexation," the democratization of the constitution, and better living conditions.

Young people who had not yet reached the age of conscription and had to worry for their lives as the war continued, refused to obey, and women, who constituted the majority of the workers in some armament factories, joined the protest movements in large numbers. Members of the forbidden New Fatherland League took part in the demonstrations and distributed leaflets.[498]

Harry Graf Kessler expected that the German government would negotiate with the workers' council, elected after the Russian model. But the politicians and military with whom he spoke on January 29 had a low opinion of negotiation. They accused the strikers of betraying the German cause. If the workers made a fuss, one should open fire. There were ten thousand reliable troop members in the capital.[499]

On January 29, one of the prepared lists of noted pacifists at the police precinct of Berlin and surroundings contained thirty-one persons—among them, in ninth place, Albert Einstein.[500] He was accused with "attempts of international character," which were punished with stricter travel regulations.[501] In fact, the "noted pacifist" was not in the streets during the strikes, nor did he meet with comrades. As he explained to the philosopher Ernst Troeltsch, when the latter complained about the lack of support for the workers on the part of the intellectuals, Einstein was currently unable to engage in any political activity.[502] While he agreed with Troeltsch, he could only follow the events from his sickbed.

The state of Einstein's health had continued to deteriorate over the past years. Unquestionably, the war with its indigestible news made him sick. And the previous year his chronic stomach and gallbladder troubles transformed into painful seizures. He had to stay in bed for weeks. The physicians prescribed him a strict diet, which could hardly be followed in the face of the food shortages in the capital and the low quality of the food that was available. His friend Zangger and south German relatives had supplied him with aid packages, sending biscuits and macaroni, rice and semolina to Berlin. But after short phases of recuperation he suffered repeatedly from further relapses.

"It's a month now that I have been lying in bed," he wrote to his older son, Hans Albert, in Zurich, at the end of January 1918. He was plagued with an obstinate stomach ulcer, because of which he had to, perhaps indefinitely, eat some sort of baby food. But that did not worry him. "I can maintain my work very well in bed, and my cousin takes excellent care of my 'bird food.'"[503]

On her own initiative his cousin Elsa Löwenthal had rented an apartment for Albert in Berlin-Schöneberg. It was in a modern corner-building, Haberlandstrasse 5, with a porter and an elevator and a separate entrance for the domestic staff. Her parents lived on the third floor, and she herself stayed on the floor above with her two daughters, Ilse and Margot, in a very grand seven-room apartment with a dining room, a parlor, and a library.[504] When the adjoining apartment suddenly became available the previous fall, she arranged an infirmary for the undemanding but nonetheless difficult patient. An interim solution, as the thirty-eight-year-old believed.

But Einstein's rest cure went on for weeks. Since the physician ordered him complete rest, he allowed himself to be taken care of by Elsa. The energy with which she stepped up to the plate to assume these duties impressed him.

As Elsa's daughter Margot later recollected, it was, after all, her mother's destiny to have to attend to everything concerning Albert—from eating to secretly smoking, which he was prohibited from doing. "He, if I may say so, remained a child."[505] He had already confessed to his lover years ago that he used a hairbrush reluctantly and a toothbrush not at all. He retired the toothbrush out of "genuine scientific considerations" because "pig bristle pierces diamonds; how then should my teeth withstand it?"[506]

Elsa let Albert get away with a great deal. She enjoyed his repartee and his sense of humor. She brought him letters almost daily and received in her parlor visitors such as the professors Planck, Born, and Emil Warburg. Einstein's entire lunch consisted of "a tiny single dish of rice cooked with milk and sugar," Planck informed their fellow physicists. But his mood was generally good and his joy of working undiminished.[507]

From his sickbed Einstein answered questions from home and abroad about his cosmological observations; reconsidered his theory of gravitational waves, smoothing away errors and pouring it into new a form; and wrote a couple of small works, though nothing of special significance, as he himself said. One finds the genuinely new only in one's youth. "The mind becomes lame, the creative powers fade, but the reputation hangs sparkling onto the calcifying shell."[508]

After he had temporarily concluded the general theory of relativity he grappled with administrative tasks. They had increased considerably since October. With his move, Haberlandstrasse 5 had become the seat of the Kaiser Wilhelm Institute for Physics.

Certainly it was a rather idiosyncratic research institute, which had neither a building nor a laboratory but quite a budget at its disposal. Einstein could distribute research grants to young physicists and lend a hand to established scientists by providing technical equipment for their laboratories. He liked this role; however, he had not considered the administrative burden that came with it.

The Prussian administrative machine had precise rules of procedure for each grant project and each expenditure, which is why Einstein applied for a secretary in December 1917. He appointed Elsa's oldest daughter, Ilse, for the approved position. The jaunty twenty-year-old dealt with this business for him for half a day three times a week and sat by "His Honor, the Director of the Kaiser Wilhelm Institute for Physics, Herr Professor Einstein himself" for dictation.[509]

Ilse had a crush on the physician and pacifist George Friedrich Nicolai, who was transferred, for disciplinary reasons, to the provincial backwater of Eilenburg near Leipzig, where she visited him on numerous occasions. More than three years earlier, Nicolai had collaborated with Einstein on the appeal "To the Europeans," which had finally been published in Switzerland as an introduction to the book *The Biology of War*, a reckoning of war based on natural science, in which Nicolai set the strategy of cooperation and mutual aid against the "struggle for existence."

Einstein himself said that he increasingly realized how insignificant everything is compared to brotherly love and humanitarianism. He deeply regretted that the language of the military had permeated everything. The new phrase "fitness training" and other similar expressions got "his bowls in uproar."[510]

Meanwhile, in western Berlin he did not witness many actual strikes, but "red Ilse"—as he called his secretary—kept him posted. On January 30, 1918, the police and the army came down hard on the demonstrators. The high command, a Prussian army head office, imposed a state of emergency on the capital. The armament companies upset by the protestors were now under strict military control. Massive numbers of striking workers were summoned to the front, and the so-called "ringleaders" were arrested. Angered by the authorities' behavior, the workers returned bit by bit to their factories. To prevent further bloodshed the strike leaders put an end to the protests.

Parallel to the show of force in its own country, the German Supreme Army Command marched the troops to the Eastern Front. The German soldiers no longer met any Russian resistance worth mentioning. It was only at the end of February that they first encountered the newly formed Red Army, which was quick to lay down its arms. The Bolsheviks had to accept an oppressive peace treaty and acknowledge the occupation of White Russia, the independence of Ukraine, Poland, the Baltic States, and Finland. With it Russia lost a big part of its territory, its industry, and its natural resources, including three quarters of the iron and coal mines. The German Reich, on the other hand, considerably increased its political sphere of influence with the new satellite states.[511]

The German workers' hopes for a lasting peace without annexations were not fulfilled. On the contrary. Ludendorff now concentrated all the available forces for a decisive battle against France. He was resolved to gamble everything in order to achieve on the Western Front what neither sides had been able to do since September: breach the opposing lines.

As Ludendorff declared in February 1918, as long as one strived for an economically strong and secured fatherland, there was no alternative to a big offensive in the west.[512] After Russia withdrew from the war, he estimated the military situation was markedly favorable. For despite the USA's entry into the war and the landing of the first American soldiers in Europe, at the beginning of 1918 the balance of power at the Western Front had not yet changed profoundly.

Armament companies such as Krupp and the AEG ran their production of artillery and shells full speed again after the strikes ended. More than 2,500 reconnaissance aircrafts and fighter-bombers were ready for operation.

Nevertheless, maintaining the supply stream was increasingly worrisome. Furthermore, Ludendorff's militant Eastern politics clouded victory prospects in the West. Around one million soldiers had to be left behind to safeguard the huge areas from the Baltic to Ukraine. Why did neither the Kaiser nor Hindenburg intervene to stop his delusion of grandeur?

Aberration of Minds and Hard Facts

As the order to attack was given in March 1918, for a short time the headlines in the *Berliner Tageblatt* resembled those from August and September 1914: "Victory at the Western Front," "Glorious progress in the big battle in France," "30,000 prisoners so far," "Bombarding Paris with long range artillery"—all with indescribable sacrifices.[513]

Einstein avoided opening the German newspapers. A few days after the offensive started, he wrote to his friend in Zurich: "politics got stuck in my stomach and rumbles there."[514]

He had never loved Germany. But in 1914 when he moved to Berlin he had not considered it possible that in view of the "general national infatuation" he would long for Germany's military defeat. "I am firmly convinced that this aberration of minds can only be cured by hard facts."[515]

In the meantime, his health "thank goodness" had become "definitely better," as Planck reported.[516] At the beginning of April, for the first time in three and a half months, he sat once again at a meeting of the Academy, where the scholars celebrated the subjugation of the eastern territories and the spreading of "German spirit" in Europe. Einstein had an allergic reaction to this jingoistic emphasis on all things German. "Your muscle-flexing under the German flag gets on my nerves," he said to a fellow professor. He would prefer to suffer than to use violence.[517]

These remarks are similar to his utterances from the onset of the war, but his tone had become bitter as a result of all the suffering and millions of war victims. Einstein simply could not grasp "that people, thoroughly decent in their personal behavior, take an entirely different attitude in relation to general affairs." Only very few individuals could free themselves from the prevailing opinions. "It seems there are no such individuals at the Academy."[518]

This comment was aimed not least at Max Planck, whom Einstein expressly mentioned in the above quoted letter to the Dutch physicist Hendrik Antoon Lorentz. Even if moderate, Planck's political stance must have especially hurt him, for there was hardly any other researcher Einstein esteemed as much as Planck. It appears from the letters they exchanged with each other during these months that Einstein tried to change his colleague's mind and persuade him with pacifist and democratic ideals. But that effort was in vain. Planck's sense of duty and his loyalty to the German Kaiser were stronger than Einstein's arguments. Up to the bitter end.

Nevertheless they remained connected to one another. They mutually recommended each other for the Nobel Prize, and Einstein prepared the celebration of Planck's sixtieth birthday, corresponding with his wife and inviting physicists from all over Germany to an official function on April 26, where he honored Planck's life work and looked back on the radical intellectual changes in physics at the beginning of the twentieth century.

At the turn of the century Planck had discovered that the energy emitted by a heated body does not escape as a continuous flow but in portions: in energy quanta. He did not at first recognize the far-reaching consequences of this discovery, partly because he considered the foundations of physics to be secured, despite all that change.

The "father of quantum theory" was conservative as a scientist as well. More than anything else he was fascinated by the unity of physics. He evaluated critical new discoveries and examined how they fit into the existing body of knowledge. Many have cut their teeth trying to cope with Planck's skepticism and resistance to all disordered, unorganized knowledge.

Einstein's freedom lay in the fact that he unfolded major physical theories—even his own—from their loose ends. He brought into light with relish every question that the existing theories omitted or ignored. Planck's formula for thermal radiation of a heated body serves as an example.

Einstein took the result seriously. He did not think of the energy quanta as mathematical artifacts. Hence he came to the conclusion that light itself consisted of light quanta or light particles with defined energy and color. This light quantum hypothesis is incompatible with those physical experiments that revealed the wave nature of light. Consequently Einstein finally came to the conclusion that the thermal radiation of a heated body has a wave character as well as a particle character. His reflections marked the beginning of the "wave-particle duality" that has shaped debates in quantum physics up to the present day.

Planck, who did not want to follow Einstein's argument, said to the assembled Academy that Einstein had overshot the target with his light quantum hypothesis. This open criticism spurred Einstein to deepen his thoughts. In the summer of 1916, he surprised Planck with new first-rate research work on the absorption and emission of radiation, which formed the foundation for the modern understanding of laser technology. "An astonishingly simple derivation of the Planck formula, I would say, *the* derivation. All entirely quantum-like."[519]

Einstein lived with open questions, not with closed systems. Almost unique among scientists, he clung to that which can easily be lost in the academic industry: the dimension of the mysterious. "The most beautiful thing we can experience is the mysterious," he once said. "It is the fundamental feeling that stands at the cradle of true art and science."[520]

Time and again Einstein broke down his colleague's reserve with his free, playful thinking. In the spring of 1918 Planck still gave preference to the wave theory of light over the light quantum hypothesis.[521] But the wave particle hypothesis no longer gave him peace. After his birthday celebration he wrote to Lorentz: "I would just like to experience in which direction quantum theory will finally go."[522] By comparison, Planck in the meantime had acquired a taste for the special and also general theory of relativity. He, Born, and Hilbert contributed greatly to the international recognition of Einstein's revolutionary discovery.

One of the many Einstein myths is that hardly any other physicist has ever understood these theories, either during the war or later. The following anecdote is one of its many variants. A journalist asked the British astronomer Arthur Eddington about Einstein's work, as he had heard that only three people had actually understood the general theory of relativity. Eddington, who had been familiar with Einstein's work since 1916, answered: "I am trying to think who could be the third one."

But in Berlin, Göttingen, Leiden, and elsewhere, researchers had long been venturing into Einstein's research field. He himself was happy with "the excitement with which the theory of relativity has been received by his colleagues, also in England and America." Neither war nor human blindness had invaded this sanctuary.[523] During the war Eddington had already prepared a British solar eclipse expedition to prove the bending of light that Einstein had predicted. This actually succeeded in 1919.

In any case, Einstein felt he was greatly indebted to his mentor Planck and other German colleagues. "I would indeed have remained a 'misunderstood genius' without the local colleagues, whom I must always bear in mind," he confided to Zangger, who would have liked to see him back in Zurich. Everything he could wish for was anticipated by his Berlin colleagues.[524]

Einstein's gratitude included Fritz Haber. The chemist used his contacts with the military to help Einstein, who at the beginning of 1918 asked his support in providing an entrance permit for a foreign researcher. But Haber neither had time to visit the sick nor shared Einstein's interests in physics. Since he had turned away from quantum physics, there had been no professional exchange of ideas between them that could have been reflected in the correspondence.

Haber was less concerned during the war with knowledge of nature than with the domination of nature. He showed his Berlin colleagues what one could achieve with research usable for military purposes, if one was determined to succeed. Haber converted a modest Kaiser Wilhelm Institute, which two years before the war had been equipped for basic research, into a large research facility committed almost exclusively to military purposes. As director, he alone could decide on the fate of the institute. The founding principle of the Kaiser Wilhelm Society had been to first choose an excellent scientist and then build an institute around him.

This principle, which has been idealized up to the present, has a few weak points, which was also revealed in the founding of the Kaiser Wilhelm Institute for Physics directed by Einstein, but in a different manner. Einstein was an outstanding researcher but by no means a good science administrator.

Haber paired scientific excellence with extraordinary leadership qualities and power ambitions. He promoted the development of poison gas in World War I with an urgency and ruthlessness that became exemplary for researchers in the Third Reich. In 1933, long after the war, Haber said: "I was one of the most powerful men in Germany.… I was more than a big military leader, more than a captain of industry. I was the founder of industries; my work was essential for the economic and military expansion of Germany. All doors stood open for me."[525]

The chemical warfare strained Haber "to the limits." He remained convinced that winning the war "was possible only with gas."[526] In the German offensive of spring 1918, the systematic firing of chemical agents that had been developed in his institute led once again to panic and to the poisoning of numerous British and French soldiers. It was also one of the reasons German troops were occasionally able to succeed in breaking the enemy's lines.

But lacking much of their military supply, the German troops were not in the position to stretch their territorial gains. Instead of pulling back now or offering peace negotiations, the Supreme Army Command persisted with its disastrous offensive tactic. In contrast to the Allies, Germany could no longer contain the losses of its soldiers. From April 1918 onward the German army shrank week by week. At the same time more and more American armed forces were landing in Europe.

Elsa or Ilse?

For the time being the home front witnessed little of the gradual exhaustion of the German troops. The military leadership controlled the flow of information. They wanted to keep the hope of a German victory alive as long as possible.

Einstein feared the war would drag on for a long time yet. He meant to publish a book with articles by established scientists from different countries "that would support and console those who have not yet lost hope in a moral development."[527] He had already secured a Swiss publisher for his project.

After careful consideration, David Hilbert declined Einstein's invitation, and he was not the only one. Such explanations amounted to "self-accusations," the mathematician warned. He recommended Einstein wait until the "hurricane of insanity" wore off. "Even your name would not protect you. Even just the word 'international' has the effect…of a red flag on our colleagues."[528]

While Einstein touted his book idea, he suffered another relapse. He was once again confined to his bed for a long time, this time because of jaundice. Suddenly he felt so weak that he canceled his yearly summer trip to see his children in Switzerland.

At any rate he could use the time to advance his divorce. After asking Mileva a second time at the beginning of the year for the dissolution of their marriage, she complied with a heavy heart. His relations with her had noticeably improved under the influence of his friends Besso and Zangger. Einstein's reaction to her nervous breakdown two years ago, as well as to her hospitalizations, had been extremely curt. After alternating diagnoses, the physicians suspected in the meantime that she suffered from scrofula.

The news from Zurich worried him greatly. His younger son, "Tete," was also constantly sick. Recently he was put up at a sanatorium in Arosa. Einstein feared his weakness and frailty were the consequences of a hereditary disease and that the boy would never again be really healthy.

Because of high health care costs, in the last year Einstein had transferred his entire annual income from the Academy to Switzerland.[529] The poor exchange rate of the mark against the Swiss franc and creeping inflation made it incrementally more difficult for him to make his alimony payment. Despite savings and his additional salary as director of the Kaiser Wilhelm Institute for physics, he had largely exhausted his finances.

In order to nevertheless meet Mileva's demands and secure the financial future of his children, he offered her a creative suggestion: since he was firmly convinced that in the coming years he would be honored with the Nobel Prize for physics, for which he had been repeatedly nominated since 1910, he would relinquish the prize money, a small fortune, to her in advance. An unusual decree in the history of the Nobel Prize.

It was a mystery to Mileva and his Swiss friends why he wanted to separate and remarry. They all knew his urge for independence and his constant need to distance himself from daily life, and reproached him for allowing himself to be roped by his relatives into a new marriage. Hadn't he spoken out against a second marriage?

By all appearances, Elsa's parents were the driving force. They expressed time and again their moral reservation against "wild marriage" (i.e., "living in sin"). Albert justified his wish for a divorce on the grounds that the marriage prospects of Elsa's daughters "would become considerably restricted by the present situation through my fault."[530] He therefore wanted to get his private affairs in order.

One then marvels all the more over a letter that was discovered in Friedrich Nicolai's estate, which reveals that Einstein had an eye on his new secretary. All of a sudden he no longer knew whom he wanted to marry: his forty-two-year-old cousin Elsa, or her daughter Ilse, who was half her age.

In mid-May 1918, Ilse returned from a visit to her beloved Nicolai in Eilenburg. The radical pacifist was famous in Berlin as a Don Juan and was known to have swung in his love affairs from mother to daughter, occasionally also the other way round. Nicolai indicated to Ilse that Albert's marriage with her would be, from his point of view, more sensible than the planned bonding with Elsa.

Shortly after Ilse's return, the issue suddenly came to a remarkably open debate. The question, first discussed half-jokingly, turned within a few minutes into a serious matter, as Ilse confided to Nicolai the next day. "Albert himself declined any decision. He is ready to marry either me or Mama. I know that A. loves me very much, perhaps as no other man will ever love me, he himself told me so

Image 10: *Traveling after the wedding: Elsa Einstein and her cousin and husband, Albert.*

yesterday." She herself, though, had never felt the slightest desire for an intimate relationship with him. "It is a different matter with him, at least lately. He himself confessed to me once how difficult it is for him to control himself."[531]

It must have been extremely amusing to Nicolai, who held an even lower opinion of middle-class conventions than his colleague, Professor Einstein, how the latter was perceived by the entire family as a trophy and what Einstein had triggered by vacillating between mother and daughter. Philistines like her grandparents were naturally horrified by the new plan, Ilse wrote. And the mother? What did Elsa say to all of this? "Temporarily—as she does not yet believe that I really take it seriously—she has allowed me a completely free choice."[532]

The letter of the twenty-year-old alternates between panic and a kind of exhilaration, which can be explained by the fact that she had spent the past years primarily among women. All the young men were at war. "We lived in a women's world," actress Marlene Dietrich said in retrospect about her years as a Berliner.[533]

Ilse, however, left no doubt that she did not want to challenge her mother's place, partly because she knew how much "all the external glamour" that would descend upon her as the wife of the most famous contemporary physicist meant to her mother. The grandparents' vanity was obviously just as big. Although Ilse herself spoke of an "intensely comical affair," she ended her long letter with a cry for help to Nicolai and added: "Please destroy this letter immediately after reading!"[534]

As Nicolai did not follow this request, we have this fine piece of evidence that Albert still did not really want to marry. He could not care less about the "formality of marriage." Was Elsa the right person for him under these conditions? Or perhaps the younger Ilse? At any rate Albert's wife would enjoy all imaginable freedom.

Nothing further is known about the direct consequences of the episode described here; it is therefore difficult to sort out Ilse's letter. In June 1918 the activist Nicolai was still harbored at the Haberlandstrasse 5 in Berlin, where he established contact with leftist circles and performed a spectacular escape: he fled Germany with a seized military aircraft and landed in Copenhagen. The event was cheered by the international press and celebrated with music at the Einstein household.[535]

A few days later, Elsa, Ilse, Margot, and Albert traveled for eight weeks in the Baltic. Albert let the sea breeze blow in his ears in the small holiday resort Ahrenshoop. Away from the capital, surrounded by his "small harem," and undisturbed by the terrible news of the war, he finally recovered from his diverse ailments. One year later he and Elsa would marry secretly and enter a bond of marriage preceded not by love declarations or oaths of loyalty, but by mutual respect and a great openness, gratitude, and pragmatism.

The End of the War and the November Revolution

After their return to the capital, things changed very rapidly. As the Allies transitioned into a counteroffensive at the Western Front, the exhaustion and the combat fatigue of the badly decimated German army manifested everywhere. The number of prisoners grew just as fast as that of the deserters. Ludendorff, the Chief of General Staff, had to finally declare bankruptcy. All at once he demanded a democratic government that would thence be capable of negotiating an armistice in accordance with the guidelines of the American president Woodrow Wilson. According to the philosopher Ernst Troeltsch, "this request for armistice was in fact a veiled capitulation."[536]

On October 26, 1918, the Kaiser dismissed Ludendorff. Three days later Wilhelm II left the capital and traveled with a large entourage to Spa, a resort town in Belgium, from whence he fought back the by now public demands for his abdication: "I have no intention of leaving the throne because of a couple hundred Jews and a thousand workers." Should the occasion arise, he would write the answer with machine guns on the pavement of Berlin. "Even if I batter my castle, order is a must."[537]

At the onset of the war, German Jews cheered the Kaiser no less than everyone else did. Many of them were quick to volunteer for the army. But their hope for reaching a social integration through the war was bitterly dashed. That the pan-German world depicted Jews as war profiteers and shirkers could surprise no one. But that the Prussian war ministry ultimately stood behind such suspicions and commanded in October 1916 a "Judenzählung"—a census to determine how many Jews subject to military duty were serving in each unit of the German army—hit the Jews in Germany as a slap in the face. They felt branded and belittled. And the longer the war lasted, the blunter the antisemitic propaganda in the Reich became. The German Kaiser's suspicions that the Jews had spread the revolution, fit into this picture.

The fate of his Russian cousin Nicholas II, who was killed in July

1918 with the whole Czarist family, caused Wilhelm II nightmares. However, by fleeing into the headquarters' lap he had left a power vacuum that revolutionary forces moved into on November 9. On the same memorable day in German history, the Berlin workers left their factories, as they had done in January, and moved with huge columns of demonstrators toward the political center.

The journalist Sebastian Haffner described their courageous protest. "They had no idea that the 'troops were no longer able to hold' and expected machine gun salvos when they arrived at the barracks and government buildings." The front rows of the endless, dull, slow marching columns that arrived from all corners of the compass carried posters with the inscriptions: "Brother, do not fire!" "The back rows carry many weapons."[538]

Instead of firing, the soldiers stationed in Berlin lay down their weapons. The Kaiser no longer had any considerable support, even among his own troops. Entire regiments followed the movement.[539]

In the city center the strikers joined with the already celebrated heroes of the revolution: the sailors who a few days earlier had put the protest in motion with open mutiny. A delegation of the sailors had flown by airship from the North Sea to Berlin-Johannisthal, where workers occupied the airfield. Another three thousand sailors arrived in Berlin by trains and motor vehicles as the day progressed.

Pressured by the events, the Imperial Chancellor Max von Baden, acting on his own authority, announced toward midday the abdication of the German Kaiser and with it the end of the monarchy. Power was up for grabs, and the Social Democrats were the first to grasp it. Friedrich Ebert immediately took the position of Imperial Chancellor; being loyal to the motto distributed in advance by his party member Philipp Scheidemann, he had to stand at the forefront of the movement in order to prevent anarchy in the Reich.

On November 9, 1918, at 14.00, from of a window of the Reichstag, Scheidemann proclaimed Germany as a republic: "The old, the ramshackle monarchy has collapsed. Militarism is finished. The Hohenzollern abdicated. Long live the new, long live the German repub-

Image. 11: *Philipp Scheidemann proclaims the republic from a window of the Reichstag.*

lic!"[540] Two hours later, Karl Liebknecht, founder of the Communist Party of Germany, who had been released from captivity just two weeks earlier, announced in front of the Berlin Palace the "free socialistic republic."

There were still shootings in front of the palace and at the royal stables, and in front of Berlin University where Albert Einstein had announced a seminar on the theory of relativity for that evening. The weekly adult education program, which he had picked up again after the long summer vacation, could not take place this time. Not because of the professor's illness, but because the entire political order had collapsed.

"November 9, 1918: Cancelled due to revolution," he entered in his lecture manuscript.[541] A week later he transitioned back into routine. "November 16: Lorentz-Transformation."[542] As if nothing else had happened.

This simultaneity of world events and the triviality of everyday life is a popular gateway for clichés: the scientist in the ivory tower indifferent toward world historic events immediately resumes his studies. In this case the representation of the unworldly researcher is further embellished through Einstein's letters.

For instance, he wrote to his mother that she should not be worried. The new Reich's leaders appeared to be equal to the task. He himself was very happy about the development of the situation and was a kind of veteran socialist among academics. "We are healthy and the Haberlandstrasse peeps into the world half curious, half anxious."[543]

That sounds—despite the joy—like middle-class smugness at a safe distance from the mass demonstrations and the revolt in the center of the city. His postcard matches the report that Ernst Troeltsch gave on the second day of the revolution as he observed the Sunday walks in the west of the city: "No elegant dress, honorable citizens, some dressed deliberately modestly. All seem subdued, like people whose destiny will be decided somewhere a long way off, and yet they are calm and comfortable that it has turned out this way.... It was written on all the faces: the salaries will continue to be paid."[544]

Einstein parked himself in this bourgeois world. The many paintings, wall hangings, and carpets in his cousin's apartment mirror a well-upholstered complacency. Some of his visitors later said that Einstein looked to them like a stranger in these surroundings.

But what happened to the other Einstein, the free-thinker and "veteran socialist"? One has to open the newspapers from the days of the revolution, the *Berliner Tageblatt*, the *Vossische Zeitung*, the *Berliner Morgenpost,* or the *Welt am Montag* in order to get on his track. Although in the past few years he had been rather secluded due to illnesses, he was now being mentioned in connection to the revolution more often than any other scientist. Suddenly he was present in all the media. Among others one learns about Einstein's political speech to more than a thousand listeners in the "Magnificent Halls of the West."

What does this mean?

The circle of engaged pacifists and democratic forward thinkers was small during the war. Until it was banned in February 1916, the New Fatherland League only had around 150 members, some with varying world views. They knew each other very well. After the ban many of them joined the founding of new organizations. Einstein, for instance, frequented the already mentioned Association of the Like-Minded, others initiated the Central Office for International Rights. Still others joined forces with the banker Richard Witting, working out the foundations for a democratic constitution. There were numerous connections between these groups, all of which were spied on by the police.

After the military collapse, the New Fatherland League reconvened in the fall of 1918. Its members turned immediately to the chancellor and his ministers and demanded the instant release of all the political prisoners and the reintroduction of the freedom of assembly, speech, and the press. At two big rallies on October 14 and 19, the League set its new political goals: a constitution of democratic and socialist spirit and "summoning a national legislative assembly with equal, secret, and direct voting rights also for women and soldiers."[545]

On the evening before the revolution, not knowing what would transpire in Berlin in the coming days, the League called for a mass rally.[546] The national assembly planned for noon on November 10 was intended to be the greatest public protest campaign since the League had come into existence. And indeed, an "unforeseeable crowd" came together at the Reichstag on this Sunday with fine fall weather. The crowd was estimated to be anywhere from many thousands up to one hundred thousand people. It is hard to believe that Einstein was not among them. Perhaps he was even on the list of speakers.

The physician Magnus Hirschfeld was the first to speak, reiterating the slogan, "Everything by the people for the people!" He had barely started to speak when "wild machine-gun fire began along the Dorotheenstrasse, interrupted by hand grenades," as the *Berliner Morgenpost* reported the next day.[547] It appears that officers loyal to the Kaiser had opened fire. The *Berliner Tageblatt* reported that after multiple shots, the revolutionary occupying force of two hundred men, who had taken hold of the Reichstag building, immediately reciprocated with gunfire from the upper floors. People flew in wild panic. "The effects of the violent fighting caused the dissolution of the planned assembly."[548]

The shattered mass rally was symptomatic of events during this second day of the revolution, in which, as Sebastian Haffner formulated so pointedly, the counter-revolution had already won the upper hand.[549] The workers returned to their factories after their revolution marches of the previous day, in order to elect a workers' council. But the well-organized social democrats pulled the strings at the center of politics. They drew the soldiers' council on their side and reached secret agreements with the military, which, in return, persisted on a special status in the ensuing republican state.

Friedrich Ebert and Philipp Scheidemann warned against a Russian-like situation. They pushed the noticeably stronger Independent Social Democrats, who were further to the left, toward a "unity of the workers' movement." In the course of the afternoon they were able to bind the opposing parties into a temporary government. On the same evening, during a turbulent meeting, they sealed together the formation of the six-head committee of the "Council of the People's Deputies."

From now on the Social Democrats controlled the political development to a massive extent. They also reached to the New Fatherland League, which enjoyed a high degree of trust in leftist circles because of its activity during the war. A few members took over political functions in the new ministries of interior, finance, and trade.

By November 11 the League had established an office in the Reichstag, a contact point for refugees, including the peace activists who had returned from exile. On that day, the war ended, a war that had cost almost ten million soldiers their lives, among them more than two million Germans. The German delegation and that of the Allies signed the armistice agreement in Compiègne.

The Berlin daily newspapers mentioned in a marginal note that the university was closed. From then on, a revolutionary student council looked after student affairs from the Reichstag. "On the same day, Einstein called me and said that…more professors, among them the rector, were interned," the physicist Max Born stated in his memoirs. Einstein feared their lives were in danger. "Since he believed he had some influence on the students, he intended to intervene; would I want to come?"[550]

Together with Born and his friend, the psychologist Max Wertheimer, Einstein left for the Reichstag building that was surrounded by a huge crowd and watched by "red soldiers." After a long wait he was able, with the help of a League member, to gain access to the building. Inside, the place was swarming with soldiers and sailors, who were lying down in the club chairs. They had piled their weapons in a pyramid.

The student council met in a small conference room and discussed the new university statutes. Einstein was sufficiently known as leftist in these circles that he and his two companions were allowed to take part in the meeting. Instead of endorsing the young people's suggestion to allow only socialist professors and students in the future, Einstein, after listening for a long while, defended academic freedom as the most valuable asset of German universities. Born remembered the astonished faces, when Einstein, whom the students considered to be on their side, refused to follow their fanaticism. The fact that he also wanted to obtain the release of the reactionary university rector Reinhold Seeberg, who during the war stood like almost no other for a victorious peace for Germany, was probably even less comprehensible to them.

After they found out that the rector and other prisoners had been handed over to the new government, the three professors proceeded to the Reich chancellor's office on Wilhelmstrasse, where they met up with journalists, trade unionists, and socialist members of parliament. "Einstein was soon recognized and greeted. Eduard Bernstein, one of his neighbors in Schöneberg, member of the New Fatherland League for many years and cofounder of the Independent Social Democrats, accompanied him to the library room, to the head of the new government. Ebert admittedly had his hands full but listened patiently to Einstein's complaint about the arrest of the rector and his colleagues and drafted the appropriate letter. "The audience thus ended and we retreated," Born recounted. The release was sealed with one pen stroke.[551]

Shortly afterward the rector and his university colleagues once again vociferously announced their revisionism. To a colleague who had encouraged them in a newspaper article, Einstein answered immediately: "The professors have provided evidence in this war that one can learn nothing from them as regards political matters but that on the contrary, it is urgently necessary for them to learn to:[552]

Shut up!"

Born's memories quoted above contradict in some details another historical source.[553] Nevertheless they impart a vivid impression of Einstein's personal spirit of optimism and his sense of justice. One wonders why the very same man fought against the professors as reactionary forces and at the same time worked for their release. Einstein felt committed to the latter, because he did not want to be as immoral as those against whose actions he had turned.

He wrote to his sister Maja on November 11 full of enthusiasm about the biggest conceivable political experience. "That I was permitted to experience that! No collapse is too big to willingly put up with it for the sake of such a marvelous compensation. The militarism and Privy Council fog are thoroughly eliminated over here."[554]

On this Monday, leaflets were circulating in Berlin with the title "Proletarians and Intellectuals Unite!" They announced a lecture by Eduard Bernstein that same evening at the Lehrervereinshaus (Teacher's Union House). It appears from the notices and an article in the *Vossische Zeitung* that professor Einstein extended his invitation to the assembly.[555]

Although he was Swiss, Einstein wanted to assert his influence on this precarious political situation and stepped onto the public stage. He made what was supposed to be his big appearance two days later at the Spichern Festival Halls. The New Fatherland League invited people there after the mass rally in front of the Reichstag was broken up. With more than one thousand listeners, the venue was filled to capacity.

The *Berliner Tageblatt* reported that "Professor Einstein spoke first in the upper hall."[556] Einstein could appear here as a veteran democrat, who did not have to learn to think differently and accept democracy. Naturally, he appealed to the Berliners to follow the Swiss example rather than the Russian.[557] According to newspaper reports he was "against the dictatorship of the proletariat" and for an immediate summoning of the National Assembly.[558]

This coincides with the content of his handwritten speech manuscript, which was found in his records but that could have been intended for another political speech. In the manuscript Einstein opposed the radical left, gave recognition to the social democratic leaders, and warned that the class tyranny of the right would be replaced by a class tyranny of the left. "Do not let yourselves be led by feelings of revenge to the disastrous view that violence can be fought with violence."[559]

His name emerged in newspapers in relation to activities outside the New Fatherland League, for instance in numerous appeals to establish the new German Democratic Party (DDP). Theodor Wolff, the chief editor of the *Berlinner Tageblatt,* and other representatives of the liberal middle class stood behind it.[560] In the election to the National Assembly the new party would win almost one fifth of the votes.

Einstein also appeared in the press numerous times as a defender of *Demokratischen Volksbund* (Democratic People's Federation). The industrialist Walther Rathenau started the party as a liberal counterweight against extreme leftists and "to give birth to a new, vigorous Reich." Along with Einstein, some big industrialists, the university rector Reinhold Seeberg, and Fritz Haber were among those advocating for the founding.[561]

How can this be explained? What is the political connection between Einstein and Haber, Seeberg, or the initiator Rathenau, who that very month had pleaded publicly for the continuation of the war and a national uprising against the impending defeat? While Einstein celebrated the days of the revolution, the world had collapsed for Haber, Seeberg, and Rathenau. Why did he now made a deal with them?

Einstein's biographers have either left this question unanswered or used his partisanship for the DDP and the People's Federation to underline his naivety regarding political questions and represent him as master of repression. Is it not typical of "Haber's friend" to also take no account of what he does not like in the programs of political parties?

The founding of the Democratic People's Federation counts as Rathenau's desperate attempt to gain a foothold in a radically changing political landscape. He wanted to be there again after the war, at the cutting edge. He defended his claim to power in the political circles of Berlin with a self-promoting and demanding tone, referring to his numerous merits. But he felt ignored by everyone, by the so-called, newly formed Council of Intellectual Workers as well as by Friedrich Ebert.

The satirist Kurt Tucholsky, the most pointed pen in the new republic, would later reprimand him in his satirical weekly magazine, *Die Weltbühne* (The World Stage). Tucholsky accused Rathenau of helping befuddle people's minds without sharing responsibility, without the honest intention of vouching for what he had preached. Disgrace was not in having erred in the war and standing for pan-Germanism even when it committed crimes. "But it was ignominy and lack of character, afterward, as soon as these convictions were no longer valid…to immediately dance to a new tune."[562]

During the chaotic revolution, Rathenau "befuddled" people's minds by engaging known scholars for his political ambitions without their knowledge. Under the founding appeal for setting up the Democratic People's Federation were names such as the art historian Wilhelm Reinhold Valentiner, who protested against being monopolized, and the sociologist Alfred Weber, who by no means defended the Democratic Federation, with which "known reactionary elements have apparently joined."[563] Nor did Einstein. The Democratic People's Federation, which had been promoted with big newspaper advertisements, dissolved into nothing after a few days. But despite Einstein's initial reluctance to support Rathenau's cause, his name in the founding appeal remained preserved in historical memory for a hundred years.

This also holds for his alleged commitment to the DDP. Einstein made it immediately clear that he also had nothing to do with Theodor Wolff's founding of the party. On November 19 the *Berliner Tageblatt* printed an opposing account as demanded by Einstein: "Prof. A Einstein is not a member of the 'Democratic Party' and declares that he has even less intention of becoming a member of the Democratic People's Federation, which had put down his name under its appeal."[564] The four-liner in the back part of the newspaper did not even attract the attention of the newspaper workers. Just two days later, Einstein's name embellished the next party appeal of the DDP, which featured prominently on the front page of the paper under the title, "The Men and Women of the New Germany!" as it had five days earlier.[565]

Einstein, as the first days of the republic show, had acquired such a high degree of fame as a researcher and pacifist that one expected his name to provide a strong political backing. His popularity relied not just on his outstanding scientific performances, but also on his persistent advocacy of freedom and democratic ideals. But he could at the time and in the future do little against the misuse of his name by colleagues and politicians for purposes he opposed.

"I enjoy the reputation of an impeccable socialist," he wrote to his friend Michele Besso at the beginning of December. "Consequently, yesterday's heroes come to me with wagging tails, believing I could thwart their fall into the void. Funny world."[566]

The New Fatherland League he had belonged to for many years and which would soon be renamed the German League for Human Rights, won many new, active allies, like the writer Heinrich Mann, the publisher and gallery owner Paul Cassirer, the diplomat and journalist Harry Graf Kessler, the painters Max Pechstein and Käthe Kollwitz, all of them belonging to the new Central Committee. Einstein wrote in the above quoted letter to Besso that something big had been achieved, which he regarded as the result of the first phase of the revolution. Admittedly, he worried about economic development and progressing inflation, but he did not let this take away his optimism. "The military religion had disappeared. I believe it will never return."[567]

He was seriously mistaken in this respect. And he was not the only one. The entire political left underestimated the aversion of the returning German soldiers to the new political and social developments. The army in particular would remain a foreign body in the fabric of the republic.[568]

In the fall of 1918 Einstein did not yet know how strong the politically motivated terror on the right would become in the very near future. Not only Karl Liebknecht and Rosa Luxemburg would fall victims to it, but many members of the New Fatherland League as well: Kurt Eisner and Gustav Landauer, Alexander Futran and Hans Paasche, while Magnus Hirschfeld and Hellmut von Gerlach survived assassination attempts against them. Geroge Friedrich Nicolai, who had returned from Copenhagen on Christmas 1918, would have his teaching permit revoked once again. In 1922 he immigrated to Argentina.

Einstein too was exposed to growing antisemitic hostility and threats by the right, which he tried to evade in part by traveling abroad. He left Berlin for longer periods of time, traveled to the USA or to Japan. The physicist, who had upended Newton's world, was received enthusiastically over there. After British researchers confirmed his general theory of relativity, his fame suddenly spread all over the globe.

By the time he was beginning to be celebrated the world over as a new Copernicus, Einstein had already long been an authority in Berlin. During the days of the revolution, his light shone brighter than ever in the capital. Jokes about him started circulating by the end of 1918. Under the title "Berliner Conversation," the *Vossische Zeitung* wrote:

"Do you know which philosopher in Berlin is quoted most often?"

"No."

"Well, Professor Einstein."

"Why?"

"Everywhere, in the city railway, elevated railway, main-line railway, one shouts on the platform before the departure: Einstei'n! (All aboard!)"[569]

Afterword

Albert Einstein has changed our understanding of nature like no other researcher.

While the fame of a physicist usually rests on being quoted by his peers, through a chain reaction Einstein was caught in the limelight of an international public. His general theory of relativity made him famous practically overnight.

That Einstein was celebrated in such a way shortly after the war is to be ascribed mostly to his epoch-making discoveries. During the years 1914 to 1918 he transformed Newton's gravitational theory, its absolute frame of reference, into a novel, not easy to comprehend perspective on space and time and the construction of our universe. If our most beautiful experience is the most mysterious, as he himself suggested, then we can see why the general theory of relativity occupied him for the rest of his life and why it has inspired human imagination up to the present day.

Einstein's international popularity after the war, however, primarily becomes understandable in view of the political situation. Einstein made a splash in Berlin as a pioneering champion of pacifism and democratic goals. In the aftermath of World War I he was promoted to a leading figure. The physicist became, in effect, the poster child of German science, which continued to be excluded from international conferences for a long time. Who but Einstein could restore its damaged reputation abroad? Abroad he was perceived as the researcher who had not allowed himself be swept up by the war euphoria of the German scholars.

But what many did not know or did not want to admit during and after the war is that Einstein went to Berlin as a Swiss citizen. Like all Swiss, a German-French war must have appeared to him as fratricide. That he behaved during the war as a committed European was for him no ground for self-adulation. He wrote to the physician Georg Friedrich Nicolai in the spring of 1918 that it would have been highly reproachful had he as Swiss taken a different stance.[570] Rather he reproached himself. He had done nothing to refurbish public opinion, he explained to Nicolai. "But I do not know if I should take my passivity amiss."[571]

Einstein here shows a modesty that singled out many other people in world wars I and II. Anyone who did anything to help others or influence public opinion believed his or her actions were natural and moreover knew how much more he or she could have done.

Since he stayed in Germany in August 1914 instead of following the exodus to Switzerland, he was, in the face of the general martial rage, first damned to "passivity." Had he ventured with a political activism à la Nicolai, who was immediately transferred for disciplinary reasons, Einstein would have been expelled from Germany. He could only hope in the first weeks of the war that the horrific episode would soon be over.

He could from then on let the war pass by him, dedicate himself completely to physics. No one would have been surprised at it. On the contrary. He was considered among colleagues an exceptional researcher. For Max Planck he already was the "new Copernicus."

Nevertheless Einstein did not feel released from care for the general good.

His entire correspondence demonstrates how much he suffered under the level of destruction and human misery and the harm to international scientific relations. Some facets of his complex personality came to light particularly during the war: his deep compassion and his intellectual independence culminating in eccentricity; his sense of social responsibility and his distressing absence as a father; his rootlessness and his solidarity with Judaism; a capacity for enthu-

Image 12: *Einstein with Planck (Einstein's left) and Nernst (Einstein's right) ten years later.*

siasm that had a contagious effect and astute perceptions that could be hurtful; his courageous, non-conforming actions and his aversion to everything military; his awareness of elitism and his modesty, sarcasm, and a deep melancholy.

The war fever of his colleagues Max Planck, Fritz Haber, and Walther Nernst, which found its clear expression when they signed the appeal "To the Civilized World" in the fall of 1914, remained incomprehensible to him. The manifesto that would significantly contribute to the boycott of German research after the war provoked his resistance. In direct response, the thirty-five-year-old supported the pacifist appeal "To the Europeans." It was the beginning of his metamorphosis.

From then on the war rapidly politicized him. He found in the New Fatherland League an opportunity to become involved in political discussions such as the debate about the goals of the war. The organization strove for a negotiated peace without annexations, and established contacts with pacifistic organizations abroad. Einstein himself emerged as a fighting intellectual.

In the meantime, the general theory of relativity ran away from him like a timid fawn that has waited too long in vain for food. As he saw his hope for a completion of his work fading, he began working feverishly on the physics material. Thus within a few weeks he succeeded in completing the theory of gravity that was founded on an entirely new understanding of space and time, the greatest individual achievement in the history of modern science.

But even this intensive work did not prevent him from writing down his opinion about the war for a lavishly designed *Patriotic Memorial Book* in the fall of 1915. In the course of World War I, Einstein changed from a physics professor committed solely to research and teaching to a representative and advocate of pacifism in Germany, visible beyond the borders of his discipline. There was a close connection between his freedom of scientific thinking and his spirited objection to the war. He had the nerve to use his own reason also in regards to political issues. He was not prepared to accept the militant nationalism in Europe, which led to an inexorable catastrophe as political reality.

Einstein's imagination freed him from intellectual fetters in the present and reached into a future in which states would join together into a League of Nations. He became the voice of hope for a peaceful co-existence of nations. That made him popular during the revolution days in Berlin and after the war mostly in the USA, where President Woodrow Wilson had long been espousing similar goals. In the years between 1914 and 1918 Einstein also laid the foundation of his later role in the German-French understanding and his membership in the International Institute for Intellectual Cooperation, the predecessor of UNESCO.

Einstein's pacifism isolated him from the scientific circles in Berlin. He lived like a drop of oil on water, as he wrote to his friend Heinrich Zangger. Further, his different approaches to life separated him from other people. "Contact is nevertheless maintained through the purely intellectual, especially through natural physics."[572]

Einstein needed fellow thinkers like theoretical physicists Max

Planck and Max Born to whom he could set forth his ideas. This was one of the principal reasons he had gone to Berlin. But even with Planck and Born, with whom he just as happily and often played music, he did not feel as personally close as he did with his friends in Switzerland and in Holland.

The "outsider" and nonconformist endured the fact that he appeared to others as "flawed"; they described his pacifism as "garish" and his political views as "naive." He faced political disagreements with autonomy and directness, as his private correspondence proves. As the war continued to be waged he consciously provoked political controversies. A culture of debate entailed unwavering tolerance. As he said afterward: "Schopenhauer's saying, 'A man can do what he wills, but not will what he wills' has been a real inspiration for me since my youth; it has been a continual consolation in the face of life's suffering and hardship, and an unfailing wellspring of tolerance."[573]

This book dedicates much attention to Einstein's misunderstood relationship with Fritz Haber. Haber endeavored to accommodate the newly arrived Berliner, whom he admired, and his family. But the two men did not have contact points in their research, nor was Haber Einstein's closest friend. The available sources indicate how ambivalent Einstein's relationship with Haber was at the onset of the war and that from 1915 onward they went their separate ways. During the remaining years of the war, there are no hints of a connection between them.

The absence of any statement by Einstein about the chemical warfare is strange. However, in contrast to the unrestricted submarine war, the deployment of chemical agents was never the object of public debate during the war. In Einstein's view, the efforts to ban poison gas after the war fell short. "It seems to me an utterly futile task to prescribe rules and limitations to war. War is not a game and cannot be carried on by rules as a game." One could only fight against war itself.[574] Did Einstein speak against the poison gas war in a personal dialogue with Haber? We do not know. But there is little ground to assume that he exercised restraint toward Haber on this point.

Einstein did not cut off contact with the "poison fanatic," as his friend Zangger called Haber, and "raving barbarian," as Einstein himself judged him. From his own experience he knew about the discrimination that his tribesman had been exposed to since childhood. With his assimilatory and military zeal, Haber, whose wife took her own life with his service pistol in May 1915, was for Einstein a tragic figure, part of the deplorable baptized Jewish Privy Council, who did not want to just integrate into German society, but rise above it. Einstein saw his fate as symbolic of "the tragedy of German Jewry, the tragedy of an unrequited love," as he wrote in 1934 to Haber's son Hermann in a condolence letter after his father's death.[575]

Theoretical physics remained throughout the entire war as Einstein's intellectual anchorage. Sometimes he lived like a hermit, drew back for days into his immeasurable world of thoughts, where he disappeared also from the eyes of the biographer; yet, he remained loyal to many questions over the years. He thought them through time and again. Since he did not come from any particular field of study, he constantly built bridges between widely separated disciplines. His thought loops and mathematical flights can hardly be shown. Every account of the fascinating origins of the general theory of relativity comes up against boundaries here.

Einstein could never understand the storm of enthusiasm that the experimental confirmation of the theory triggered after World War I. A year later he wrote in a letter to the physicist Max von Laue: "If I have learned one thing in a long life of pondering, it is that we are much further away from a deep insight into the elementary processes than most of our contemporaries believe…so that noisy celebrations hardly correspond to the actual situation."[576]

The German physicist von Laue had a different view on the matter: "Posterity will apply another yardstick. It will not ask how far removed such a man was from the goals he had set himself, but how much knowledge he added to the previously found treasure."[577] And how much he championed so that this knowledge would become useful for the peaceful coexistence of people.

Acknowledgments

I would like to thank everyone who supported me with this book, above all Barbara Wenner, Christian Koth, and Hanser Publishers, Alexander Zock, Stefan Klein, and Jörg Resag, the workers at the Berlin State Library, the archive of the Max Planck Society, as well as the researchers who published Einstein's manuscripts and letters, and commented on and opened up his work. The Max Planck Institute for the History of Science in Berlin, particularly Jürgen Renn, Hans-jakob Ziemer, Giuseppe Castagnetti, and Urs Schoepflin, deserve my thanks for their support during my stay as "journalist in residence" at the Institute and during a later guest residence. All failures in this book are naturally mine.

— *Thomas de Padova*, Berlin, July 2015

Translator's Note

Special thanks to:

Thomas de Padova, author; *Marianne Ward*, translation editor; *James Munves*, Bunim & Bannigan founder.

— *Michal Schwartz*, Prince Edward Island, Canada, March 2018

Endnotes

1 Jost, “Einstein und Zürich,”19.
2 Einstein in Stachel et al., *Collected Papers* Vol. 5, 103.
3 *Deutsche Hochschulstimmen*, 351.
4 *Collected Papers* Vol. 8, 410.
5 Ibid., 562.
6 Sloterdjik, *Zur Welt kommen*, 129.
7 Trischler, *Luft- und Raumfahrtforschung in Deutschland*, 45–46.
8 *Intelligenzblatt der Stadt Bern*, July 14, 1913.
9 Flugsport, *Illustrierte Flugtechnische Zeitschrift*, 510.
10 *Intelligenzblatt der Stadt Bern*, July 14, 1913.
11 Ibid.
12 Walter, *Bider, der Flieger*, 219.
13 Ibid.
14 Kafka, *Ein Landarzt und andere Drucke zu Lebzeiten*, 318.
15 Dienel, *Herrschaft über die Natur*?, 140.
16 *Intelligenzblatt der Stadt Bern*, July 14, 1913.
17 Kristen & Treder, *Albert Einstein in Berlin*, 97.
18 *Collected Papers* Vol. 5, 467.
19 Grüning, *Ein Haus für Albert Einstein*, 185.
20 Ibid., 175–6.
21 Kirsten & Treder, 97.
22 Planck, *Acht Vorlesungen über theoretische Physik*, 117.
23 Kirsten & Treder, 95.
24 Ibid., 96.
25 *Collected Papers* Vol. 5, 505.
26 Pais, *Raffiniert ist der Herrgott*, 240.
27 *Collected Papers* Vol. 5, 505.
28 Schilpp, *Albert Einstein als Philosoph*, 16.
29 *Collected Papers* Vol. 5, 432–3.
30 Ibid., 499.
31 Reiser, *Albert Einstein*, 75.
32 Seelig, *Albert Einstein und die Schweiz*, 101.
33 *Collected Papers* Vol. 5, 510–1.
34 Reiser, 90.
35 *Collected Papers* Vol. 5, 543.
36 Ibid., 536–7.
37 Ibid., 537.
38 Ibid., 456.
39 Ibid., 518.
40 Ibid., 469.
41 Castagnetti et al., *Einstein in Berlin*.
42 *Collected Papers* Vol. 5, 538.
43 Kirsten & Treder, 101–2.
44 Kirchhoff, *Die akademische Frau. Gutachten*, 320–1.

45 Ibid., 256f.
46 Popovic, *In Albert's Shadow*, 4.
47 *Collected Papers* Vol. 1, 254.
48 Ibid., 248.
49 Ibid., 253f.
50 Fölsing, A. 2001, 41.
51 *Collected Papers* Vol. 5, 345.
52 Fölsing, *Nobel-Frauen*, 41.
53 Goldsmith, *Marie Curie*, 168.
54 Beuys, *Die neuen Frauen*, 297.
55 *Collected Papers* Vol. 5, 544.
56 Sloterdijk, *Scheintod im Denken*, 51.
57 Highfield & Carter, *Die geheimen Leben des Albert Einstein*, 163.
58 Popovic, *In Albert's shadow*, 16.
59 *Collected Papers* Vol. 5, 573–4.
60 Fölsing, *Albert Einstein*, 593.
61 *Collected Papers* Vol. 5, 544.
62 Reid, *Marie Curie*, 158–9.
63 Curie, *Marie Curie*, 250.
64 Moszkowski, *Einstein*, 10.
65 Einstein, «Zum gegenwärtigen Stande des Gravitationsproblems," 1249–50.
66 Keisinger, *Unzivilisierte Kriege im zivilisierten Europa?*, 44.
67 *Freie Presse*, August 11, 1913.
68 Piper, *Nacht über Europa*, 292–3.
69 Clark, *Die Schlafwandler*, 320.
70 Keisinger, *Unzivilisierte Kriege im zivilisierten Europa?*, 121–2.
71 Clark, *Die Schlafwandler*, 376.
72 *Collected Papers* Vol. 5, 508.
73 Clark, *Die Schlafwandler*, 360.
74 Keisinger, *Unzivilisierte Kriege im zivilisierten Europa ?*, 124–5.
75 Ibid.
76 *Prager Tageblatt*, August 11, 1913.
77 Clark, *Die Schlafwandler*, 361.
78 Musil, *Der Mann Ohne Eigenschafte*, 9.
79 *Berliner Morgenpost*, March 27, 1914.
80 März, *Ernst Ludwig Kirchner*, 4.
81 Fürst, *Emil Rathenau*, 95.
82 Posener, *Berlin auf dem Weg zu einer neuen Architektur*, 32.
83 Fürst, *Emil Rathenau*, 115.
84 Clark, *Preussen. Aufstieg und Niedergang*, 672.
85 *Berliner Taageblatt*, March 30, 1914.
86 Schmitt, *Als die Oldtimer flogen*, 163.
87 Ibid., 102f.
88 Von Suttner, *Die Barbarisierung der Luft*, 5.
89 Schmitt, 130.
90 Neffe, *Einstein*, 115.
91 *Collected Papers* Vol. 5, 565.

92 Ibid., 585.
93 Zott, *Fritz Haber in seiner Korrespondenz*, 75.
94 Hoffmann, *Einstein's Berlin*, 12.
95 *Collected Papers* Vol. 8, 13.
96 Hahn, *Mein Leben*, 106.
97 *Collected Papers* Vol. 5, 574.
98 Clark, *Preussen*, 684 f.
99 Seelig, *Alber Einstein. Mein Weltbild*, 13.
100 Girardet, *Jüdische Mäzene*, 37.
101 Ibid., 8.
102 Kollros, "Erinnerungen eines Kommilitonen," 30.
103 *Vossische Zeitung*, April 16, 1914.
104 Ibid.
105 Ibid.
106 Einstein, "Zur Elektrodynamik bewegter Körper."
107 *Vossische Zeitung*, April 16, 1914.
108 Pössel, *Das Einstein-Fenster*, 70.
109 *Vossische Zeitung*, April 16, 1914.
110 Einstein, "Zur Elektrodynamik bewegter Körper," 893.
111 Elias, Über die Zeit, VIII–IV.
112 Einstein, "Zur Elektrodynamik bewegter Körper," 893.
113 Sexl & Schmidt, *Raum – Zeit – Relativität*, 33.
114 *Vossische Zeitung*, April 16, 1914.
115 Quoted in Promies, *Georg Christoph Lichtenberg*, 224.
116 *Collected Papers* Vol. 8, 50.
117 Ibid., 49.
118 Ibid., 47.
119 *Freie Presse*, July 10, 1914.
120 *Collected Papers* Vol. 8, 17.
121 Frank, *Einstein. Sein Leben und seine Zeit*, 387.
122 *Collected Papers* Vol. 8, 28.
123 Quoted in Frisé, *Robert Musil. Gesammelte Werke*. Vol. II, 1007.
124 Scheel, *Verhandlungen der Deutschen Physikalischen Gesellschaft*, 1914.
125 *Collected Papers* Vol. 5, 34.
126 Gehrke, *Kritik der Relativitätstheorie*, 34–35.
127 Gehrke, *Die Massensuggestion der Relativitätstheorie*, 5.
128 Wazeck, *Einsteins Gegner*, 133.
129 *Collected Papers* Vol. 5, 555.
130 Gehrke, *Kritik der Relativitätstheorie*, 19.
131 Einstein, "Die Relativitäts-Theorie," 12.
132 Wazeck, *Einsteins Gegner*, 124.
133 Gehrke, *Kritik der Relativitätstheorie*, 35.
134 *Collected Papers* Vol. 8, 29.
135 *Berliner Tageblatt*, July 3, 1914.
136 Planck, "Erwiderung an Hrn. Einstein," 742–3.
137 Ibid.
138 Ibid.

139 Kirsten & Treder, *Albert Einstein in Berlin,* 104.
140 Scheel, *Verhandlungen der Deutschen Physikalischen Gesellschaft,* 457f, 512–3.
141 Ibid., 735, 820–1.
142 Münkler, *Der grosse Krieg,* 36.
143 Röhl, *Wilhelm II,* 1131.
144 Leonhard, *Die Büchse der Pandora,* 91–92.
145 *Collected Papers* Vol. 8, 41.
146 *Collected Papers* Vol. 5, 572–3.
147 *Collected Papers* Vol. 8, 48–49.
148 Ibid.
149 Ibid., 1032–3.
150 Ibid., 44.
151 Ibid., 45.
152 Fischer-Dücklemann, *Die Frau als Hausärztin,* 245–66.
153 Ibid.
154 Beuys, *Die neuen Frauen,* 174.
155 *Collected Papers* Vol. 8, 1032–3.
156 Ibid, 47.
157 Hölzle, *Quellen zur Entstehung des Ersten Weltkriegs,* 398–9.
158 Institute for Marxismus-Leninismus, *Dokumente und Materialien,* 492–3.
159 Kruse, "Welche Wendung durch des Weltkrieges Schickung," 116.
160 *Berliner Tageblatt,* July 19, 1914.
161 Kuczynski, *Der Ausbruch des Ersten Weltkrieges,* 57.
162 Afflerbach, *Kaiser Wilhelm II,* 130.
163 Hölzle, *Quellen zur Entstehung des Ersten Weltkriegs,* 420.
164 Ibid., 424.
165 Ibid., 433.
166 Münkler, *Der grosse Krieg,* 82–83.
167 Hölzle, *Quellen zur Entstehung des Ersten Weltkriegs* 429–30.
168 Stoltzenberg, *Fritz Haber,* 230.
169 *Collected Papers* Vol. 8, 49–50.
170 Schulmann, *Seelenverwandte,* 113.
171 *Collected Papers* Vol. 8, 50.
172 Ibid., 52.
173 *Collected Papers* Vol. 8, 562.
174 Leonhard, *Die Büchse der Pandora,* 102.
175 Reichsarchiv, *Der Weltkrieg 1914 bis 1918,* 32–33.
176 Knipping, *Eisenbahn im Krieg,* 40.
177 Ibid.
178 178 Von Moltke, *Gesammelte Schriften und Denkwürdigkeiten,* 38–39.
179 Heinze, *Räder rollen für den Krieg,* 43.
180 *Collected Papers* Vol. 8, 56.
181 Ibid., 112.
182 Frisé, *Robert Musil,* 1020–1.
183 Verhey, *Der "Geist von 1914" und die Erfindung der Volksgemeinschaft,* 167–8.
184 Rürup, "Es entspricht nicht dem Ernste der Zeit," 182.
185 Verhey, *Der "Geist von 1914,"* 162.

186 Ibid., 192.
187 *Collected Papers* Vol. 13, 747.
188 Heilbron, *Max Planck*, 94.
189 Wilde, *Walther Rathenau*, 79.
190 Röhl, *Wilhelm II*, 1174–5.
191 Basler, "Zur politischen Rolle der Berliner Universität," 201.
192 Ibid., 182.
193 *Deutsche Hochschulstimmen*, 351.
194 Johann, *Innenansicht eines Krieges*, 67.
195 Wilde, *Walther Rathenau*, 82.
196 Zott, *Wilhelm Ostwald und Walther Nernst*, 77.
197 Mendelssohn, *Walther Nernst und seine Zeit*, 112–3.
198 Benrabi, "Die Kulturmission der Schweiz," 210.
199 Schulmann, *Seelenverwandte*, 112–3.
200 Lipp, *Pazifismus im Ersten Weltkrieg*, 15.
201 *Deutsche Hochschulstimmen*, 395.
202 Kühlem, *Carl Duisberg (1861–1935)*, 195–6.
203 Johann, *Innenansicht eines Krieges*, 61.
204 Rolland, *Das Gewissen Europas*, 47.
205 Ungern & Sternberg, *Der Aufruf.*
206 *Berliner Tageblatt*, October 4, 1914.
207 Ibid.
208 Szöllösi-Janze, *Fritz Haber*, 259.
209 Brock, "Wissenschaft und Militarismus," 667–8.
210 Reinbothe, *Deutsch als internationale Wissenschaftssprache*, 422.
211 Rolland, *Das Gewissen Europas*, 400.
212 Mac-Leod, "Mobilmachung der Forscher," 3.
213 Kox, *The Scientific Correspondence of H. A. Lorentz*, 395.
214 Brocke, "Wissenschaft und Militarismus," 686.
215 *Collected Papers* Vol 8, 63.
216 Kox, *Scientific Correspondence*, 446.
217 Glasser, *Wilhelm Conrad Roentgen*, 119.
218 Tollmien, "Der "Krieg der Geister in der Provinz," 187.
219 Ungern & Sternberg, *Der Aufruf*, 64–65.
220 Kox, *Scientific Correspondence*, 446.
221 Ibid., 427–8.
222 Schulmann, *Seelenverwandte*, 116–7.
223 Ibid., 256f.
224 Frank, *Einstein. Sein Leben und seine Zeit*, 198.
225 Ibid.
226 Ibid.
227 Riemer, *Die Postüberwachung im Deutschen Reich*, 6.
228 Nicolai, *Nachrichtendienst*, 41.
229 Schulmann, *Seelenverwandte*, 116–117.
230 *Collected Papers* 1998, Vol 8, 45.
231 Schulmann, *Seelenverwandte*, 172.
232 Vierhaus & vom Brocke, *Forschung im Spannungsfeld*, 177.

233 Reiser, *Albert Einstein*, 138.
234 Zuelzer, *Der Fall Nicolai*, 144.
235 Lipp, *Pazifismus im Ersten Weltkrieg*, 26–27.
236 Vom Brocke, "Wissenschaft und Militarismus," 683.
237 Einstein quoted in Schulmann, *Seelenverwandte*, 116.
238 Ibid.
239 Haber, *Mein Leben mit Fritz Haber*, 90.
240 Haffner, *Der Verrat*, 24–25.
241 Sösemann & Frölich, *Theodor Wolff*, 211–2.
242 Haffner, *Der Verrat*, 21.
243 *Collected Papers* Vol. 8, 85.
244 Hahn, *Mein Leben*, 119.
245 Schmidt-Ott, *Erlebtes und Erstrebtes*, 124.
246 Max-Planck-Gessellschat Archive, Va Abt., Rep. 0005, no. 856.
247 Stern, "Freunde im Widerspruch. Haber und Einstein," 523.
248 Frank, *Einstein. Sein Leben und seine Zeit*, 251.
249 Wilde, *Walther Rathenau*, 19–20.
250 Einstein, "Wie ich Zionist wurde," 352.
251 *Collected Papers* Vol. 8, 18.
252 Seelig, *Albert Einstein. Mein Weltbild*, 170–1.
253 Ibid., 153.
254 *Collected Papers* Vol. 8, 53.
255 Jaenicke, *100 Jahre Bunsen-Gesellschaft 1894–1994*, 62.
256 Stoltzenberg, *Fritz Haber*, 155.
257 Szöllösi-Janze, *Fritz Haber*, 178–9.
258 Mittasch, *Geschichte der Ammoniaksynthese*, 116.
259 Jaenicke, *100 Jahre Bunsen-Gesellschaft 1894–1994*, 49.
260 Leitner, *Der Fall Clara Immerwahr*, 148.
261 Willstätter, *Aus meinem Leben*, 203.
262 Leitner, *Der Fall Clara Immerwahr*, 191.
263 Haber, *Aus Leben und Beruf*, 13.
264 Szöllösi-Janze, *Fritz Haber*, 315.
265 Ekstrand, "Award Ceremony Speech," 321–2.
266 Szöllösi-Janze, *Fritz Haber*, 314–5.
267 Schulmann, *Seelenverwandte*, 340.
268 Stoltzenberg, *Fritz Haber*, 396.
269 Fölsing, *Albert Einstein*, 752.
270 Max-Planck-Gesellschaft, Archive, III, Abt., Rep. 98, no. 58.
271 Born, *Albert Einstein, Hedwig und Max Born*, 39.
272 Ibid., 40.
273 Münkler, *Der grosse Krieg*, 362–3.
274 Baumann, *Giftgas und Salpeter*, 258–9.
275 Sommerfeld, "Zum siebzigsten Geburtstag Albert Einsteins," 144.
276 Baumann, *Giftgas und Salpeter*, 286.
277 Szöllösi-Janze, *Fritz Haber*, 272.
278 Max-Planck-Gesellschaft Archive, Va Abt., Rep. 0005, no. 1479.
279 Ernst, *Lise Meitner an Otto Hahn*, 26.

280 Münkler, *Der grosse Krieg*, 289.
281 Scheel, *Verhandlungen der Deutschen Physikalischen Gesellschaft*, 41.
282 Bartel & Huebener, *Walther Nernst, Pioneer of Physics and of Chemistry*, 256.
283 Leonhard, *Die Büchse der Pandora*, 182.
284 Baumann, *Giftgas und Salpeter*, 294.
285 Martinez, *Der Gaskrieg 1914–1918*, 18.
286 Mendelssohn, *Walther Nernst und seine Zeit*, 113.
287 Schulmann, *Seelenverwandte*, 116.
288 *Collected Papers* 1998, Vol 8, 85.
289 Schulmann, *Seelenverwandte*, 117.
290 *Collected Papers* Vol. 8, 113.
291 Archiv der Berlin Brandenburgischen Akademie der Wissenschaften, PAW II (1912-1945), Signatur II–V, 90–94 & 133.
292 Max-Planck-Gesellscaft Archive, Va Abt., Rep. 0005, no. 860.
293 *Collected Papers* Vol. 10, 275.
294 Martinez, *Der Gaskrieg 1914–1918*, 13–14.
295 Ibid., 20.
296 Ibid., 20 f.
297 Baumann, *Giftgas und Salpeter*, 343.
298 Born, *Mein Leben*, 261.
299 Martinez, *Der Gaskrieg 1914–1918*, 42.
300 Hahn, *Mein Leben*, 117 f.
301 Ernst, *Lise Meitner an Otto Hahn*, 41.
302 Max-Planck-Gesellschaft Archive, Va Abt., Rep. 0005, no. 1480.
303 Ibid.
304 Ibid., no. 1470.
305 Schulmann, *Seelenverwandte*, 129.
306 Ibid., 124.
307 Kox, *Scientific Correspondence*, 427–8.
308 Ibid.
309 Münkler, *Der grosse Krieg*, 295.
310 Kühlem, *Carl Duisberg (1861–1935)*, 222.
311 Lehmann-Russbüldt, *Der Kampf der Deutschen Liga für Menschenrechte*, 48–49.
312 Münkler, *Der grosse Krieg*, 291–3.
313 Eisenbeiss, *Die bürgerliche Friedensbewegung in Deutschland*, 136.
314 Scheer, *Die Deutsche Friedensgesellschaft (1892–1933)*, 248.
315 Grundmann, *Einsteins Akte*, 49.
316 Ibid., 48–49.
317 *Collected Papers* Vol. 8, 103.
318 Lehmann-Russbüldt, *Der Kampf der Deutschen Liga für Menschenrechte*, 30.
319 Leonhard, *Die Büchse der Pandora*, 294.
320 Niedhart, *Gustav Mayer*, 363.
321 Erdmann, *Kurt Riezler*, 270.
322 Kühlem, *Carl Duisberg (1861–1935)*, 234–5.
323 Hahn, *Mein Leben*, 119.
324 Niedhart, *Gustav Mayer*, 370.

325 *Collected Papers* Vol. 8, 129.
326 Ibid.
327 Rolland, *Das Gewissen Europas*, 696.
328 Ibid.
329 Ibid., 700.
330 Ibid., 697–8.
331 Eisenbeiss, *Die bürgerliche Friedensbewegung in Deutschland*, 139.
332 *Collected Papers* Vol. 6, 211.
333 *Das Land Goethe*, 1916, 30.
334 Grundmann, *Einsteins Akte*, 49.
335 Eckert & Märker, *Arnold Sommerfeld*, 501.
336 *Collected Papers* Vol. 8, 177–8.
337 Fölsing, *Albert Einstein*, 418.
338 Einstein, Über die spezielle und allgemeine Relativittstheorie, 42.
339 Einstein & Infeld, *Die Evolution der Physik*, 168.
340 Ibid., 164.
341 Einstein, Über die spezielle und allgemeine Relativittstheorie, 41.
342 Ibid., 42.
343 Einstein & Infeld, *Die Evolution der Physik*, 46.
344 Born, *Mein Leben*, 234.
345 Einstein, "How I Created the Theory of Relativity," 45–46.
346 Pössel, *Das Einstein-Fenster*, 105.
347 Einstein, "Über das Relativitätsprinzip und die aus demselben gezogenen Folgerungen," 454.
348 Ibid., 461.
349 *Collected Papers* Vol. 5, 317.
350 Eckert & Märker, *Arnold Sommerfeld*, 510.
351 Einstein, "Über den Einfluss der Schwerkraft auf die Ausbreitung des Lichtes," 493.
352 Staude & Hofmann, "Sonnenforschung in Potsdam," 110.
353 Einstein & Infeld, *Die Evolution der Physik*, 260.
354 Chou et al., "Optical Clocks and Relativity," 1630–31.
355 Von Laue, *Gesammelte Schriften und Vorträge*, Band II, 23.
356 Kollros, "Erinnerungen eines Kommilitonen," 27.
357 *Collected Papers* Vol. 5, 505.
358 Seelig, *Albert Einstein. Mein Weltbild*, 228.
359 De Padova, *Das Weltgeheimnis*, 227.
360 Quoted in Heintz, *Die Innenwelt der Mathematik*, 48–49.
361 Einstein, "Geometrie und Erfahrung," 2.
362 Ibid., 1.
363 Ibid., 3.
364 Wussing, *6000 Jahre Mathematik*, 458.
365 Schirrmacher, "Theoretiker zwischen mathematischer und experimenteller Physik," 43.
366 Corry, "David Hilbert between Mechanical and Electromagnetic Reductionism 1910–1915,"489f.
367 *Collected Papers* Vol. 8, 147.

368 Tollmien, Der "Krieg der Geister" in der Provinz, 146.
369 Busse, *Engagement oder Rückzug*?, 240.
370 Howard & Norton, "Out of the labyrinth?," 39.
371 *Collected Papers* Vol. 8, 181.
372 Schulmann, *Seelenverwandte*, 192.
373 Collected Papers Vol. 13, 265.
374 Sauer & Majer, *David Hilbert's Lectures on the Foundations of Physics*, 167.
375 Ibid., 108.
376 *Collected Papers* Vol. 8, 91.
377 Ibid., 113.
378 Ibid., 146.
379 Ibid., 168.
380 Rolland, *Das Gewissen Europas*, 624–5.
381 *Collected Papers* Vol. 8, 178.
382 Einstein, "Zur allgemeinen Relativitätstheorie," 778.
383 Ibid.
384 Ibid., 779.
385 Einstein, "Zur allgemeinen Relativitätstheorie (Nachtrag)," 799.
386 *Collected Papers* Vol. 8, 195f.
387 Ibid., 199.
388 Einstein, "Erklärung der Perihelbewegung des Merkur aus der allgemeinen Relativitätstheorie," 831–2.
389 *Collected Papers* Vol. 8, 202.
390 Ibid., 201.
391 Renn & Sauer, "Einsteins Züricher Notizbuch," 865–6.
392 Renn, *Auf den Schultern von Riesen und Zwergen*, 281.
393 Einstein, "Die Feldgleichungen der Gravitation," 847.
394 *Collected Papers* Vol. 8, 217.
395 Seelig, *Albert Einstein. Mein Weltbild*, 228–9.
396 De Padova, *Leibniz, Newton und die Erfindung der Zeit*, 238–9.
397 *Collected Papers* Vol. 8, 205.
398 Ibid., 222.
399 Ibid., 291.
400 Einstein, "Näherungsweise Integration der Feldgleichungen der Gravitation."
401 *Collected Papers* Vol. 8, 366.
402 Ibid., 225.
403 Scheel, *Verhandlungen der Deutschen Physikalischen Gesellschaft*, 261.
404 *Collected Papers* Vol. 8, 411.
405 Born, *Physik im Wandel meiner Zeit*, 195.
406 Born, "Albert Einstein ganz privat," 36.
407 Born, *Physik im Wandel meiner Zeit*, 193.
408 *Collected Papers* Vol. 8, 223.
409 Max-Planck-Gesellschaft Archive, Va Abt., Rep. 0005, no. 1470.
410 Frank, *Einstein. Sein Leben und seine Zeit*, 188.
411 Ibid.
412 Sloterdijk, *Scheintod im Denken*, 49–50.
413 Schulmann, *Seelenverwandte*, 309–10.

414 Ibid., 180.
415 Ibid.
416 Seelig, *Albert Einstein. Leben und Werk eines Genies unserer Zeit*, 259–60.
417 Döring, *Der Weimarer Kreis*, 56.
418 Heilbron, *Max Planck. Ein Leben für die Wissenschaft 1858–1947*, 262–3.
419 Eisenbeiss, *Die bürgerliche Friedensbewegung in Deutschland*, 140.
420 Gülzow, "Der Bund," 234.
421 Grundmann, *Einsteins Akte*, 51.
422 *Collected Papers* Vol. 8, 636.
423 Einstein, "Ernst Mach,"104.
424 *Collected Papers* Vol. 8, 104.
425 Ernst, *Lise Meitner an Otto Hahn*, 64.
426 Gülzow, "Der Bund," 234.
427 Schulmann, *Seelenverwandte*, 184.
428 Lehmann-Russbüldt, *Der Kampf der Deutschen Liga für Menschenrechte*, 46; *Collected Papers*, 1998, Vol. 8, 134.
429 Schulmann, *Seelenverwandt*, 203.
430 Seelig, *Albert Einstein. Mein Weltbild*, 79.
431 Meinecke, *Autobiographische Schiften*, 254.
432 Schulmann, *Seelenverwandte*, 192.
433 *Collected Papers* Vol. 8, 399.
434 Leonhard, *Die Büchse der Pandora*, 445.
435 Holitscher, *Mein Leben in dieser Zeit*, 114.
436 Leonhard, *Die Büchse der Pandora*, 445.
437 Born, *Mein Leben*, 241.
438 Scheel, *Verhandlungen der Deutschen Physikalischen Gesellschaft*, 297.
439 Ibid., 318–9.
440 Schwabe, *Wissenschaft und Kriegsmoral*, 95.
441 Wehler, *Deutsche Gesellschaftsgeschichte 1914–1949*, 71.
442 Mendelssohn, *Walther Nernst und seine Zeit*, 125.
443 Born, *Mein Leben*, 246.
444 Schulmann, *Seelenverwandte*, 186.
445 Schwabe, *Wissenschaft und Kriegsmoral*, 104.
446 Nernst, "Der Krieg und die deutsche Industrie," 1207.
447 Ibid.
448 Schwabe, *Wissenschaft und Kriegsmoral*, 97.
449 Holl, "Die Vereinigung Gleichgesinnter," 374.
450 Born, *Mein Leben*, 256.
451 Ibid.
452 Martinez, *Der Gaskrieg 1914–1918*, 70.
453 Baumann, *Giftgas und Salpeter*, 386.
454 Hahn, *Mein Leben*, 122.
455 Ibid., 132.
456 Max-Planck-Gesellschaft Archive, Va Abt., Rep. 0005, no. 858.
457 Ibid.
458 Ibid., Nt. 963.
459 Leonhard, *Die Büchse der Pandora*, 296.

460 Martinez, *Der Gaskrieg 1914–1918*, 79.
461 *Collected Papers* Vol. 8, 386.
462 Ibid.
463 Einstein, "Kosmologische Betrachtungen zur allgemeinen Relativitätstheorie."
464 Einstein, Über die spezielle und die allgemeine Relativitätstheorie, 71–72.
465 Einstein, "Kosmologische Betrachtungen zur allgemeinen Relativitätstheorie," 143.
466 Ibid., 144.
467 Einstein, Über die spezielle und die allgemeine Relativitätstheori, 71–72.
468 Ibid.
469 Born, *Mein Leben*, 234.
470 Einstein, "Kosmologische Betrachtungen zur allgemeinen Relativitätstheorie," 152.
471 Renn, *Auf den Schultern von Riesen und Zwergen*, 293.
472 *Collected Papers* Vol. 8, 411.
473 Flugsport, *Illustrierte Flugtechnische Zeitschrift für das gesamte Flugwesen* 1917, 93–94.
474 Münkler, *Der grosse Krieg*, 453.
475 Einstein, "Elementare Theorie der Wasserwellen und des Fluges," 509.
476 *Collected Papers* Vol. 13, 256.
477 *Interavia*, Professor Einsteins "Leichtsinn," 684.
478 Ibid.
479 Illy, *The Practical Einstein*, 72–73.
480 *Interavia*, Professor Einsteins "Leichtsinn," 684.
481 Ibid.
482 Ibid.
483 *Collected Papers* Vol. 8, 577.
484 *Berliner Tageblatt*, October 1918.
485 Schmitt, *Als die Oldtimer flogen*, 180.
486 Leonhard, *Die Büchse der Pandora*, 862.
487 Inspektion des Flugzeugswesens (Inspection of Aircraft System), 92.
488 *Collected Papers* Vol. 8, 588.
489 Archiv der BBAW, PAW II (1912–1945), Signatur II–V, 93, Blatt 12, 22 & 133, Blatt 96–97.
490 Max-Planck-Gesellschaft Archive, III. Abt., Rep. 98, no. 36.
491 Szöllösi-Janze, *Fritz Haber*, 403 f.; Haber, *Mein Leben mit Fritz Haber*, 113.
492 *Collected Papers* Vol. 8, 465.
493 *Berliner Tageblatt*, December 25, 1917.
494 *Collected Papers* Vol. 8, 506.
495 Ibid.
496 Kessler, *Das Tagebuch*, 262.
497 Ibid., 268.
498 Gülzow, "Der Bund," 373.
499 Kessler, *Das Tagebuch*, 262.
500 Grundmann, *Einsteins Akte*, 63.
501 Fölsing, *Albert Einstein*, 464.

502 *Collected Papers* Vol. 8, 636.
503 Ibid., 614.
504 Hoffmann, *Einsteins Berlin*, 23.
505 Grüning, *Ein Haus für Albert Einstein*, 457.
506 *Collected Papers* Vol. 5, 570–1.
507 Kox, *Scientific Correspondence*, 495.
508 *Collected Papers* Vol. 8, 849.
509 Ibid., 613.
510 Schulmann, *Seelenverwandte*, 283.
511 Leonhard, *Die Büchse der Pandora*, 813–4.
512 Münkler, *Der grosse Krieg*, 677.
513 *Berliner Tageblatt*, March 27, 1918.
514 Schulmann, *Seelenverwandte*, 297.
515 *Collected Papers* Vol. 8, 505.
516 Kox, *Scientific Correspondence*, 495.
517 *Collected Papers* Vol. 8, 663.
518 Ibid., 430.
519 Ibid., 329.
520 Seelig, *Albert Einstein. Mein Weltbild*, 13–14.
521 Kox, *Scientific Correspondence*, 500.
522 Ibid., 502.
523 Schulmann, *Seelenverwandte*, 209.
524 Ibid., 256–7.
525 Stolzenberg, *Fritz Haber*, 620.
526 Max-Planck-Gesellscaft Archive, III, Abt., Rep. 98, no. 27.
527 *Collected Papers* Vol. 8, 736.
528 Ibid., 745.
529 Fölsing, *Albert Einstein*, 472.
530 *Collected Papers* Vol. 8, 667.
531 Ibid., 769–70.
532 Ibid.
533 Wenzel, “Schöneberg voran!,” 163.
534 *Collected Papers* Vol. 8, 769–70.
535 Zuelzer, *Der Fall Nicolai*, 229.
536 Troeltsch, *Spektator-Briefe*, 8.
537 Röhl, *Wilhelm II*, 1242.
538 Haffner, *Der Verrat. 1918 / 1919*, 77.
539 *Vossische Zeitung*, November 1918.
540 Clark, *Preussen*, 704.
541 *Collected Papers* Vol. 7, 90.
542 Ibid.
543 *Collected Papers* Vol. 8, 945.
544 Troeltsch, *Spektator-Briefe*, 24.
545 Lehmann-Russbüldt, *Der Kampf der Deutschen Liga für Menschenrechte*, 80–81.
546 Gülzow, “Der Bund,” 410.
547 *Berliner Morgenpost*, November 11, 1918.

548 *Berliner Tageblatt*, November 11, 1918.
549 Haffner, *Der Verrat*, 97–98.
550 Born, *Mein Leben*, 257–8.
551 Ibid.
552 *Collected Papers* Vol. 8, 945.
553 Holitscher, *Mein Leben in dieser Zeit*, 162; Goenner, *Einstein in Berlin*, 114–5.
554 *Collected Papers* Vol. 8, 945.
555 *Vossische Zeitung*, November 10, 1918.
556 *Berliner Tageblatt*, November 14, 1918.
557 *Collected Papers* Vol. 7, 123.
558 *Berliner Tageblatt*, November 14, 1918.
559 *Collected Papers* Vol. 7, 123.
560 *Berliner Tageblatt*, November 16, 1918; *Vossische Zeitung*, November 16, 1918.
561 *Vossische Zeitung*, November 18, 1918; *Berliner Tageblatt*, November 19, 1918; *Berliner Morgenpost*, November 19, 1918.
562 *Die Weltbühne*, May 29, 1919.
563 Schölzel, *Walther Rathenau*, 264; *Berliner Tageblatt*, November 19, 1918.
564 Ibid.
565 *Berliner Tageblatt*, November 21, 1918.
566 *Collected Papers* Vol. 8, 959.
567 Ibid., 958.
568 Clark, *Preussen*, 715.
569 *Vossische Zeitung*, December 22, 1918.
570 *Collected Papers* Vol. 8, 759.
571 Ibid.
572 Schulmann, *Seelenverwandte*, 208.
573 Seelig, *Albert Einstein. Mein Weltbild*, 10.
574 Nathan & Norden, *Albert Einstein. Über den Frieden*, 109.
575 Max-Planck-Gesellschaft Archive, III, Abt., Rep. 98, no. 58.
576 Von Laue, *Gesammelte Schriften und Vorträge*, Bd. III, 228.
577 Ibid., 229.

Bibliography

Afflerbach, Holger. *Kaiser Wilhelm II. als Oberster Kriegsherr im Ersten Weltkrieg. Quellen aus der militärischen Umgebung des Kaisers*. München: De Gruyter, 2005.

Bartel, Hans Georg. "Ein Geheimrat im Militärdienst." *Physik Journal* 13, 7 (2014).

Bartel, Hans Georg, and Rudolf P. Huebener. *Walther Nernst, Pioneer of Physics and of Chemistry*. London: World Scientific, 2007.

Basler, Werner. "Zur politischen Rolle der Berliner Universität im ersten imperialistischen Weltkrieg 1914 bis 1918." *Wissenschaftliche Zeitschrift der Humboldt-Universität zu Berlin*, Gesellschafts-und sprachwissenschaftliche Reihe. 10 (1961): 181–203.

Baumann, Timo. *Giftgas und Salpeter*. Düsseldorf, 2008.

Benrabi, I. "Die Kulturmission der Schweiz." *Internationale Monatsschrift für Wissenschaft, Kunst und Technik* 10, 10 (1916): 4.

Berlin-Brandenburgische Akademie der Wissenschaften, Archive. PAW II (1912–1945), Signatur II-V, 90–94, 133.

Beuys, Barbara. *Die neuen Frauen – Revolution im Kaiserreich 1900–1914*. München: Carl Hanser, 2014.

Bohnke-Kollwitz, Jutta, ed. *Käthe Kollwitz. Die Tagebücher.* Berlin: Siedler, 1989.

Born, Hedwig. "Albert Einstein ganz privat." In *Helle Zeit – Dunkle Zeit*, edited by Carl Seelig, 35–39. Zürich: Vieweg +Teubner, 1956.

Born, Max. *Albert Einstein, Hedwig und Max Born. Briefwechsel 1916–1955*. München: Edition Erbrich, 1969.

———. *Mein Leben*. München: Nymphenburger Verlagshandlung, 1975. [*Recollections of a Nobel Laureate*. New York: Scribner, 1978.]

———. *Physik im Wandel meiner Zeit*. Braunschweig: Springer, 1983. [*Physics in My Generation*. Oxford: Pergamon Press, 1956.]

———. *Die Relativitätstheorie Einsteins*. Heidelberg: Springer, 2003. [*Einstein's Theory of Relativity*. Mineola, NY: Dover Publications, 1961.]

Brenner, Wolfgang. *Walther Rathenau. Deutscher und Jude*. München: Piper, 2005.

Busse, Detlef. *Engagement oder Rückzug? Göttinger Naturwissenschaften im Ersten Weltkrieg*. Göttingen: Universitätsverlag, 2008.

Castagnetti, Giuseppe, Hubert Goenner, et al. *Einstein in Berlin. Wissenschaft zwischen Grundlagenkrise und Politik*. Berlin: MPI für Bildungsforschung, 1994.

Chou, Chin-Wen, D. B. Hume, T. Rosenband, and D. J. Wineland. "Optical Clocks and Relativity." *Science* 329, 5999 (Sept. 24, 2010): 1630–3.

Clark, Christopher. *Preussen. Aufstieg und Niedergang 1600–1947*. München: Pantheon, 2008. [*Iron Kingdom: The Rise and Downfall of Prussia, 1600–1947*. Translated by Norbert Juraschitz. Cambridge: Belknap Press, 2009.]

———. *Die Schlafwandler. Wie Europa in den Ersten Weltkrieg zog.* München: Deutsche Verlags Anstalt, 2013. [*The Sleepwalkers. How Europe Went to War in 1914*. Translated by Norbert Juraschitz. London: Allan Lane, 2012.]

Corry, Leo. "David Hilbert between Mechanical and Electromagnetic Reductionism 1910–1915." *Archive for History of Exact Sciences* 53, 6 (1999): 489–527.

Curie, Eve. *Madame Curie: Eine Biographie*. Frankfurt am Main: Fischer, 1994. [*Madame Curie. A Biography.* Translated by Vincent Sheean. New York: Doubleday, 1937.]

Das Land Goethe 1914–1916. Ein vaterländisches Gedenkbuch. Berlin: Goethe Bund, 1916.

De Padova, Thomas. *Das Weltgeheimnis. Kepler, Galilei und die Vermessung des Himmels.* München: Piper, 2009.

———. *Leibniz, Newton und die Erfindung der Zeit*. München: Piper, 2013.

Deutsche Hochschulstimmen 33, 1914.

Dienel, Hans-Liudger. *Herrschaft über die Natur? Das Naturverständnis deutscher Ingenieure 1871–1914*. Stuttgart: Verlag für Geschichte der Naturwissenschaften und der Technik, 1992.

Döring, Herbert. *Der Weimarer Kreis. Studien zum Bewusstsein verfassungstreuer Hochschullehrer in der Weimarer Republik.* Meisenheim am Glan: Hain, 1975.

Eckert, Michael, and Karl Märker, eds. *Arnold Sommerfeld: Wissenschaftlicher Briefwechsel*. Vol. 1, 1892–1918. München: GNT-Verlag, 2000.

Einstein, Albert. "Zur Elektrodynamik bewegter Körper." ["On the Electrodynamics of Moving Bodies."] *Annalen der Physik* 17 (1905): 891–921.

———. "Über das Relativitätsprinzip und die aus demselben gezogenen Folgerungen." ["On the Relativity Principle and the Consequences Drawn from It."] *Jahrbuch der Radioaktivität und Elektronik* (1907): 411–69.

———. "Über den Einfluss der Schwerkraft auf die Ausbreitung des Lichtes." ["On the Influence of Gravitation on the Propagation of Light."] *Annalen der Physik* 35 (1911): 485–98.

———. "Die Relativitäts-Theorie." ["The Theory of Relativity."] *Vierteljahrsschrift der Naturforschenden Gesellschaft in Zürich* 56 (1912): 1–14.

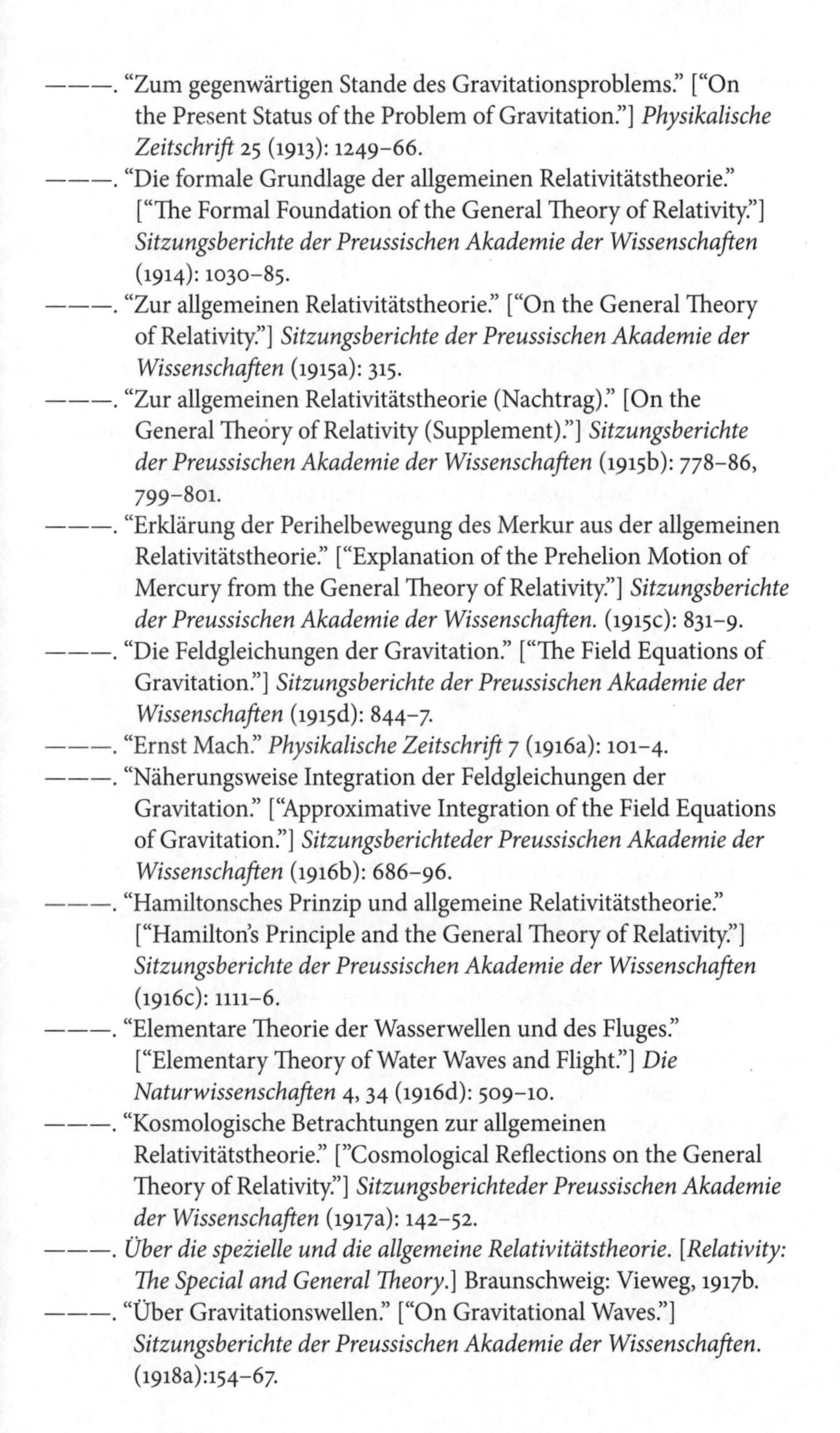

———. “Zum gegenwärtigen Stande des Gravitationsproblems.” [“On the Present Status of the Problem of Gravitation.”] *Physikalische Zeitschrift* 25 (1913): 1249–66.

———. “Die formale Grundlage der allgemeinen Relativitätstheorie.” [“The Formal Foundation of the General Theory of Relativity.”] *Sitzungsberichte der Preussischen Akademie der Wissenschaften* (1914): 1030–85.

———. “Zur allgemeinen Relativitätstheorie.” [“On the General Theory of Relativity.”] *Sitzungsberichte der Preussischen Akademie der Wissenschaften* (1915a): 315.

———. “Zur allgemeinen Relativitätstheorie (Nachtrag).” [On the General Theory of Relativity (Supplement).”] *Sitzungsberichte der Preussischen Akademie der Wissenschaften* (1915b): 778–86, 799–801.

———. “Erklärung der Perihelbewegung des Merkur aus der allgemeinen Relativitätstheorie.” [“Explanation of the Prehelion Motion of Mercury from the General Theory of Relativity.”] *Sitzungsberichte der Preussischen Akademie der Wissenschaften.* (1915c): 831–9.

———. “Die Feldgleichungen der Gravitation.” [“The Field Equations of Gravitation.”] *Sitzungsberichte der Preussischen Akademie der Wissenschaften* (1915d): 844–7.

———. “Ernst Mach.” *Physikalische Zeitschrift* 7 (1916a): 101–4.

———. “Näherungsweise Integration der Feldgleichungen der Gravitation.” [“Approximative Integration of the Field Equations of Gravitation.”] *Sitzungsberichteder Preussischen Akademie der Wissenschaften* (1916b): 686–96.

———. “Hamiltonsches Prinzip und allgemeine Relativitätstheorie.” [“Hamilton’s Principle and the General Theory of Relativity.”] *Sitzungsberichte der Preussischen Akademie der Wissenschaften* (1916c): 1111–6.

———. “Elementare Theorie der Wasserwellen und des Fluges.” [“Elementary Theory of Water Waves and Flight.”] *Die Naturwissenschaften* 4, 34 (1916d): 509–10.

———. “Kosmologische Betrachtungen zur allgemeinen Relativitätstheorie.” [”Cosmological Reflections on the General Theory of Relativity.”] *Sitzungsberichteder Preussischen Akademie der Wissenschaften* (1917a): 142–52.

———. *Über die spezielle und die allgemeine Relativitätstheorie.* [*Relativity: The Special and General Theory.*] Braunschweig: Vieweg, 1917b.

———. “Über Gravitationswellen.” [“On Gravitational Waves.”] *Sitzungsberichte der Preussischen Akademie der Wissenschaften.* (1918a):154–67.

———. “Kritisches zu einer von Hrn. De Sitter gegebenen Lösung der Gravitationsgleichungen.” [“Critical Comment on a Solution to the Gravitational Equations by Mr. de Sitter.”] *Sitzungsberichte der Preussischen Akademie der Wissenschaften* (1918b): 270–2.

———. “Geometrie und Erfahrung.” [“Geometry and Experience.”] *Sitzungsberichte der Preussischen Akademie der Wissenschaften* (1921a): 1–8.

———. “Wie ich Zionist wurde.” [“How I Became a Zionist.”] *Jüdische Rundschau* 49, 2 (Juni 1921b): 351–2.

———. “How I Created the Theory of Relativity.” *Physics Today* 8 (1982): 45–47.

———. *Über die spezielle und allgemeine Relativittstheorie*. Braunschweig: Vieweg, 1992. [*Relativity: The Special and General Theory*. Translated by Robert W. Lawson. Overland Park: Digireads.com, 2012.]

Einstein, Albert, and Leopold Infeld. *Die Evolution der Physik*. Köln: Anaconda, 2014. [*The Evolution of Physics*. New York: Simon & Schuster, 1938.]

Eisenbeiss, Wilfried. *Die bürgerliche Friedensbewegung in Deutschland während des Ersten Weltkriegs*. Frankfurt am Main: Peter Lang, 1980.

Ekstrand, Åke Gerhard. “Award Ceremony Speech.” In *Nobel Foundation, Nobel Lectures, Chemistry 1901–1902*. Amsterdam: Elsevier Publishing Company, 1966.

Elias, Norbert. *Über die Zeit*. Frankfurt am Main: Suhrkamp, 1998. [*Time: An Essay*. Translated by Edmund Jephcott. New York: Blackwell, 1993.]

Erdmann, Karl Dietrich, ed. *Kurt Riezler, Tagebücher, Aufsätze, Dokumente*. Göttingen: Vandenhoeck & Ruprecht, 1972.

Ernst, Sabine. *Lise Meitner an Otto Hahn. Briefe aus den Jahren 1912 bis 1924*. Stuttgart: Wissenschaftlich Verlagsgeselschaft, 1992.

Fischer-Dückelmann, Anna. *Die Frau als Hausärztin*. Stuttgart: Süddeutsches Verlags Institute, 1913.

Flugsport. *Illustrierte Flugtechnische Zeitschrift für das gesamte Flugwesen*. Frankfurt am Main, 1909–1944.

Fölsing, Albrecht. *Albert Einstein*. Frankfurt am Main: Suhrkamp, 1995. [*Albert Einstein: A Biography*. Translated by Ewald Osers. New York: Penguin, 1997.]

Fölsing, Ulla. *Nobel-Frauen. Naturwissenschaftlerinnen im Porträt*. München: C. H. Beck, 2001.

Frank, Phillip. “Einsteins Stellung zur Philosophie.” *Deutsche Beiträge* 2 (1949).

———. *Einstein. Sein Leben und seine Zeit*. Braunschweig: Springer, 1979. [*Einstein: His Life and Times*. Translated by George Rosen. Boston: De Capo Press, 2002.]

Fraunholz, Uwe. *Motorphobia. Anti-automobiler Protest in Kaiserreich und Weimarer Republik*. Göttingen: Vandenhoeck & Ruprecht, 2000.

Fries, Helmut. *Die grosse Katharsis*. Konstanz: Verlag am Hockgraben, 1994.

Frisé, Adolf, ed. *Robert Musil. Gesammelte Werke*. Vol. II, Hamburg: Rowohlt, 1983.

Fuchs, Margot. *Georg von Arco (1869–1940) – Ingenieur, Pazifist, Technischer Direktor von Telefunken*. Berlin: GNT, 2004.

Fürst, Arthur. *Emil Rathenau. Der Mann und sein Werk*. Berlin: Vita, 1915.

Gehrke, Ernst. *Kritik der Relativitätstheorie*. Berlin: Herman Meuser, 1924a.

——. *Die Massensuggestion der Relativitätstheorie. Kulturhistorisch-Psychologische Dokumente*. Berlin: Meuser, 1924b.

Girardet, Cella Margaretha. *Jüdische Mäzene für die Preussischen Museen zu Berlin*. Berlin: Hänsel-Hohenhausen, 1997.

Giulini, Domenico. *Am Anfang war die Ewigkeit*. München: C. H. Beck, 2004.

Glasser, Otto. *Wilhelm Conrad Roentgen und die Geschichte der Entdeckung der Röntgenstrahlung*. Berlin: Springer, 1931.

Glatzer, Ruth. *Das wilhelminische Berlin*. Berlin: Siedler, 1997.

Goenner, Hubert. *Einstein in Berlin*. München: C. H. Beck, 2005.

Goldsmith, Barbara. *Marie Curie. Die erste Frau der Wissenschaft*. München: Piper, 2010. [*Obssesive Genius: The Inner World of Marie Curie*. Translated by Sonja Hauser. New York: Norton and Company, 2005.]

Grundmann, Siegfried. *Einsteins Akte*. Berlin: Springer, 1998. [*The Einstein Dossiers*. Berlin: Springer, 2005.]

Grüning, Michael. *Ein Haus für Albert Einstein. Erinnerungen, Briefe, Dokumente*. Berlin: Verlag der Nation, 1990.

Gülzow, Erwin. "Der Bund. Neues Vaterland." Dissertation, Humboldt University Berlin, 1969.

Haber, Charlotte. *Mein Leben mit Fritz Haber*. Düsseldorf: Econ, 1970.

Haber, Fritz. *Aus Leben und Beruf*. Berlin: Springer, 1927.

Haber, Ludwig Fritz. *The Poisonous Cloud: Chemical Warfare in the First World War*. Oxford: Oxford University Press, 1986.

Haenisch, Konrad. *Die deutsche Sozialdemokratie in und nach dem Weltkriege*. Berlin: C. A. Schwetschke, 1916.

Haffner, Sebastian. *Der Verrat. 1918 / 1919 – als Deutschland wurde, wie es ist*. Berlin: Verlag 1900, 1994.

———. *Geschichte eines Deutschen*. Stuttgart: Deutsche Verlags-Anstalt, 2000. [*Defying Hitler: A Memoir*. Translated by Oliver Pretzel. New York: Farrar, Strauss and Giroux, 2000.]

Hahn, Otto. *Mein Leben*. München, 1968. [*My Life: The Autobiography of a Scientist*. Translated by Ernst Kaiser and Eithne Wilkins. New York: Herder and Herder, 1970.]

Heilbron, John L. *Max Planck. Ein Leben für die Wissenschaft 1858–1947*. Stuttgart: S. Hirzel, 1988. [*Dilemmas of an Upright Man: Max Planck and the Fortunes of German Science*. Harvard University Press, 2000.]

———. *Max Planck*. Stuttgart: S. Hirzel, 2006.

Heintz, Bettina. *Die Innenwelt der Mathematik*. Wien: Springer, 2000.

Heinze, Dieter. *Räder rollen für den Krieg. Die militärische Nutzung der Eisenbahn von den frühen Anfängen bis 1989*. Leipzig: Engelsdorfer, 2008.

Henning, Eckart, and Marion Kazemi. *Dahlem – Domäne der Wissenschaft*. Berlin: Archiv zur Geschichte Max-Planck-Ges., 2009.

Hermann, Armin. *Einstein – Der Weltweise und sein Jahrhundert*. München: Piper, 1994.

Highfield, Roger, and Paul Carter. *Die geheimen Leben des Albert Einstein*. München: Mairx, 1994. [*The Private Lives of Albert Einstein*. New York: St. Martin's Griffin, 1994.]

Hoffmann, Dieter. *Einsteins Berlin. Auf den Spuren eines Genies*. Weinheim: Willey-VCH, 2006. [*Einstein's Berlin: In the Footsteps of a Genius*. Baltimore: Johns Hopkins University Press, 2013.]

Holitscher, Arthur. *Mein Leben in dieser Zeit*. Potsdam: Kipenheuer, 1928.

Holl, Karl. "Die Vereinigung Gleichgesinnter. Ein Berliner Kreis pazifistischer Intellektueller im Ersten Weltkrieg." Archiv für Kulturgeschichte 54 (1972): 364–84.

Holton, Gerald. *Einstein, die Geschichte und andere Leidenschaften*. Braunschweig: Veweg, 1998. [*Einstein, History, and Other Passions*. Copernicus Books, 1995.]

Hölzle, Ervin, ed. *Quellen zur Entstehung des Ersten Weltkriegs. Internationale Dokumente 1901–1914*. Darmstadt: Wiss. Buchges, 1995.

Horgan, John. *An den Grenzen des Wissens*. Translated by Thorsten Schmidt. München: Fischer, 1996. [*The End of Science*. New York: Broadway Books, 1996.]

Howard, Don, and John Norton. "Out of the Labyrinth? Einstein, Hertz, and the Göttingen Answer to the Hole Argument." In *The Attraction of Gravitation: New Studies in the History of General Relativity*, edited by J. Earman, Michel Janssen, and John D. Norton. Boston: Birkhäuser, 1993.

Illy, Joseph. *The Practical Einstein*. Baltimore: Johns Hopkins University Press, 2012.

Inspektion des Flugzeugwesens. *Geschichte der deutschen Flugzeugindustrie.* Berlin, 1918.

Institut für Marxismus-Leninismus beim Zentralkomitee der Sozialistischen Einheitspartei Deutschlands, ed. *Dokumente und Materialien zur Geschichte der deutschen Arbeiterbewegung.* Vol. IV. Berlin: Dietz, 1975.

Interavia. Professor Einsteins "Leichtsinn." Vol. 10. 9, Frankfurt am Main, 1955.

Jaenicke, Walther. *100 Jahre Bunsen-Gesellschaft 1894–1994.* Darmstadt: Steinkopf, 1994.

Johann, Ernst, ed. *Innenansicht eines Krieges, Deutsche Dokumente 1914–1918.* Frankfurt am Main: Büherglide Gutenberg, 1968.

Jost, Res. "Einstein und Zürich. Zürich und Einstein." In *Vierteljahrsschrift der Naturforschenden Gesellschaft in Zürich* Vol. 124, edited by Eugen A. Thomas, 7–23. Zürich, 1979.

Kafka, Franz. *Ein Landarzt und andere Drucke zu Lebzeiten.* Frankfurt am Main: Fischer, 1994. [*The Tales of Franz Kafka.* Translated by Alessandro Baruffi. Philadelphia: Literary Joint Press, 2016.]

Kanitscheider, Bernulf. *Von der mechanistischen Welt zum kreativen Universum.* Darmstadt: WBG, 1993.

Keisinger, Florian. *Unzivilisierte Kriege im zivilisierten Europa? Die Balkankriege und die öffentliche Meinung in Deutschland, England und Irland 1876–1913.* Paderborn: Schöningh, 2008.

Kessler, Harry Graff. *Das Tagebuch,* Sixth Vol., 1916–1918. Stuttgart: Cotta, 2006. [*Journey to the Abyss: The Diaries of Count Harry Kessler, 1880–1918.* Translated by Laird Easton. New York: Vintage, 2013.]

Kirchhoff, Arthur. *Die akademische Frau. Gutachten hervorragender Universitätsprofessoren, Frauenlehrer und Schriftsteller über die Befähigung der Frau zum wissenschaftlichen Studium und Berufe.* Berlin: H. Steinitz, 1897.

Kirsten, Christa and Hans Jürgen Treder. *Albert Einstein in Berlin (1913–1933), Darstellungen und Dokumente.* Berlin: Akademie, 1979.

Knipping, Andreas. *Eisenbahn im Krieg.* München: Geramond, 2005.

Köhler, Henning. "Berlin in der Weimarer Republik." *Geschichte Berlins, Vol. II.* Berlin: 2002, 797–925.

Kollros, Louis. "Erinnerungen eines Kommilitonen." In *Helle Zeit – Dunkle Zeit,* edited by Carl Seelig, 17–31. Zürich: Europa, 1956.

Kox, Anne. J., ed. *The Scientific Correspondence of H. A. Lorentz,* Vol.1, New York: Springer, 2008.

Krumeich, G. and M. R. Lepsius. *Max Weber. Briefe 1918–1920.* Tübingen: Paul Siebeck, 2012.

Kruse, W. "Welche Wendung durch des Weltkrieges Schickung. – Die SPD und der Beginn des Ersten Weltkrieges." In *August 1914. Ein Volk zieht in den Krieg*, 115–126. Berliner Geschichtswerkstatt. Berlin: Nishen, 1989.

Kuczynski, Jürgen. *Der Ausbruch des Ersten Weltkrieges und die deutsche Sozialdemokratie*. Berlin: Akademie, 1957.

Kühlem, Kordula. *Carl Duisberg (1861–1935). Briefe eines Industriellen*. München: Oldenbourg, 2012.

Lehmann-Russbüldt, Otto. *Der Kampf der Deutschen Liga für Menschenrechte vormals Bund Neues Vaterland für den Weltfrieden*. Berlin: Hensel & Company, 1927.

Leitner, Gerit von. *Der Fall Clara Immerwahr*. München: C. H. Beck, 1993.

Leonhard, Jörg. *Die Büchse der Pandora. Geschichte des Ersten Weltkriegs*. München: C. H. Beck, 2014. [*Pandora's Box: A History of the First World War*. Translated by Patrick Camiller. Cambridge: Belknap Press, 2018.]

Lipp, Karlheinz. *Pazifismus im Ersten Weltkrieg*. Herbolzheim: Centaurus, 2004.

Livio, Mario. *Ist Gott ein Mathematiker?* München: C. H. Beck, 2010. [*Is God a Mathematician?* New York: Simon & Schuster, 2010.]

Mac-Leod, Roy. "Mobilmachung der Forscher." *Physik Journal* 13, 7 (2014): 3.

Martinez, Dieter. *Der Gaskrieg 1914–1918*. Bonn: Bernard & Graefe, 1996.

März, R. *Ernst Ludwig Kirchner. Potsdamer Platz 1914*. Berlin, 2000.

Max-Planck-Gesellschaft Archive, III. Abt., Rep. 98, no. 36.

———. III,Abt., Rep. 98, no. 58.

———. Va Abt., Rep. 0005, no. 858.

———. Va Abt., Rep. 0005, no. 1470.

———. Va Abt., Rep.0005, no. 1479.

———. Va Abt., Rep 0005, no.1480.

Meinecke, Friedrich. *Autobiographische Schiften*. Stuttgart: De Gruyter Odenburg, 1969.

Mendelssohn, Kurt. *Walther Nernst und seine Zeit*. Weinheim: Physik, 1976. [*The World of Walther Nernst: The Rise and Fall of German Science, 1864–1941*. Pittsburgh: University of Pittsburgh Press, 1973.]

Mittasch, Alwin. "Der Stickstoff als Lebensfrage." *Deutsches Museum, Abhandlungen und Berichte* 13 (1941): 1–34.

———. *Geschichte der Ammoniaksynthese*. Weinheim: Chemie, 1951.

Moszkowski, Alexander. *Einstein. Einblicke in seine Gedankenwelt*. Berlin: Hoffman und Kampe, 1921.

Mudry, Anna. *Galileo Galilei – Schriften, Briefe, Dokumente*. Berlin: Rütten & Loening, 1987.

Münkler, Herfried. *Der grosse Krieg. Die Welt 1914–1918*. Berlin: Rowohlt, 2014.

Musil, Robert. *Der Mann ohne Eigenschaften*. Edited by Adolf Fris. Hamburg: Rowohlt, 1978. [*The Man Without Qualities*. Translated by Sophie Wilkins and Bertun Pike. New York: Alfred A. Knopf, 1995.]

Nathan, Otto, and Heinz Norden, eds. *Albert Einstein. Über den Frieden*. Bern: Peter Lang, 1975. [*Einstein on Peace*. London: Methuen and Co., 1963.]

Nationalgalerie Berlin. *Ernst Ludwig Kirchner 1880–1938*. Berlin, 1980.

Neffe, Jürg. *Einstein. Eine Biographie*. Hamburg: Rowohlt, 2005. [*Einstein: A Biography*. Baltimore: Johns Hopkins University Press, 2009.]

Nernst, Walther. "Der Krieg und die deutsche Industrie." *Internationale Monatsschrift für Wissenschaft, Kunst und Technik* 10, 10 (1916).

Nicolai, Walter. *Nachrichtendienst, Presse und Volksstimmung im Weltkrieg*. Berlin: E. S. Mittler und Sohn, 1920.

Niedhart, Gottfried, ed. *Gustav Mayer. Als deutsch-jüdischer Historiker in Krieg und Revolution 1914–1920*. München: Oldenburg, 2009.

Pais, Abraham. *"Reffiniert is der Herrgott..." Albert Einstein. Eine Wissenschaftliche Biographie*. Wiesbaden: Vieweg Braunschweig, 1986. [*Subtle is the Lord: The Science and the Life of Albert Einstein*. New York: Oxford University Press, 1982.]

Piper, Ernst. *Nacht über Europa. Kulturgeschichte des Ersten Weltkriegs*. Berlin: Propyläen, 2013.

Planck, Max. *Acht Vorlesungen über theoretische Physik*. Leipzig: S. Hirzel, 1910. [*Eight Lectures on Theoretical Physics*. New York: Dover Publications, 1997.]

———. "Erwiderung an Hrn. Einstein." Sitzungsberichte der Königlich Preussischen Akademie der Wissenschaften. Vol. II, Berlin, 1914, 739–44.

Popovic, Milan. *In Albert's Shadow: The Life and Letters of Mileva Maric, Einstein's First Wife*. Baltimore: Johns Hopkins University Press, 2003.

Posener, Julius. *Berlin auf dem Weg zu einer neuen Architektur, Das Zeitalter Wilhelms II*. München: Prestel, 1979.

Pössel, Markus. *Das Einstein-Fenster. Eine Reise in die Raumzeit*. Hamburg: Hoffman und Campe, 2005.

Promies, Wolfgang, ed. *Georg Christoph Lichtenberg, Aphorismen, Schriften, Briefe*. München: Zweitausendeins, 1991.

Reichsarchiv. *Der Weltkrieg 1914 bis 1918. Das deutsche Feldeisenbahnwesen. Vol. 1: Die Eisenbahnen zu Kriegsbeginn*. Berlin: E. S. Mitler und Sohn, 1928.

Reid, Robert. *Marie Curie*. New York: Signet, 1974.
Reinbothe, Roswitha. *Deutsch als internationale Wissenschaftssprache und der Boykott nach dem Ersten Weltkrieg*. Frankfurt am Main: Peter Lang, 2006.
Reiser, Anton. *Albert Einstein*. New York: Albert & Charles Boni, 1930.
Renn, Jürgen and T. Sauer. "Einsteins Züricher Notizbuch. Die Entdeckung der Feldgleichungen der Gravitation im Jahre 1912." *Physikalische Blätter* 52, 9 (1996): 865–72.
Renn, Jürgen. *Albert Einstein – Ingenieur des Universums. Hundert Autoren für Einstein*. Berlin: Wiley-VCH, 2005a.
———, ed. *Einstein's Annalen Paper: The Complete Collection 1901–1922*. Weinheim: Wiley-VCH, 2005b.
———. *Auf den Schultern von Riesen und Zwergen*. Weinheim: Wiley-VCH, 2006a.
———. *The Genesis of General Relativity*. Dordrecht: Springer, 2006b.
Riemer, Karl-Heinz. *Die Postüberwachung im Deutschen Reich durch Postüberwachungsstellen 1914–1918*. Neue Schriftenreihe Poststempelgilde "Rhein-Donau" 109. Düsseldorf, 1987.
Rife, Patricia. *Lise Meitner. Ein Leben für die Wissenschaft*. Hildesheim: Claassen,1992. [*Lise Meitner and the Dawn of a Nuclear Age*. Basel: Birkhäuser, 1999.]
Röhl, John. *Wilhelm II. – Der Weg in den Abgrund 1900–1941*. München: C. H. Beck, 2009.
Rolland, Romain. *Das Gewissen Europas. Tagebuch der Kriegsjahre 1914–1919*. Berlin: Rütten & Leoning, 1963.
Rowe, David E. and Robert Schulmann, eds. *Einstein on Politics*. Princeton: Princeton University Press, 2007.
Rürup, Ingeborg. "Es entspricht nicht dem Ernste der Zeit, dass die Jugend müssig gehe." In *August 1914. Ein Volk zieht in den Krieg*, 181–93. Berliner Geschichtswerkstatt. Berlin: Nishen, 1989.
Russell, Bertrand. *Das ABC der Relativitätstheorie*. Translated by Kurt Grelling. München: Nymphenburger Verlagshandlung, 1970. [*The ABC of Relativity*. New York: Harper & Brothers, 1925.]
Sauer, Tilman, and Ulrich Majer, eds. *David Hilbert's Lectures on the Foundations of Physics*. Heidelberg: Springer, 2009.
Scheel, Karl, ed. *Verhandlungen der Deutschen Physikalischen Gesellschaft, 16. bis 19. Jahrg*. Berlin: 1914–1918.
Scheer, Firedrich Karl. *Die Deutsche Friedensgesellschaft (1892–1933). Organisation, Ideologie, Politische Ziele*. Frankfurt am Main: Haag und Herchen, 1983.

Schilpp, Paul Arthur, ed. *Albert Einstein als Philosoph und Naturforscher*. Braunschweig: Vieweg + Teubner, 1979. [*Albert Einstein Philosopher-Scientist*. Northwest University: Library of Living Philosophers, 1949.]

Schirrmacher, Arne. "Theoretiker zwischen mathematischer und experimenteller Physik – zu Max Plancks Stil physikalischen Argumentierens." In *Max Planck und die moderne Physik*, edited by Dieter Hoffmann, 35–48. Heidelberg: Springer, 2010.

Schirrmacher, Arne. "Die Physik im Grossen Krieg." *Physik Journal* 13, 7 (2014): 43.

Schlick, Moritz. "Die philosophische Bedeutung des Relativitätsprinzips." In *Zeitschrift für Philosophie und philosophische Kritik* 159 (1915), edited by Michael Stöltzner and Thomas Uebel, 129–75. *Wiener Kreis*. Hamburg: Brill, 2006.

Schmidt-Ott, Friedrich. *Erlebtes und Erstrebtes. 1860–1950*. Wiesbaden: Steiner, 1952.

Schmitt, Günther. *Als die Oldtimer flogen. Die Geschichte des Flugplatzes Berlin-Johannisthal*. Berlin: Transpress, 1987.

Schölzel, Christian. *Walther Rathenau*. Paderborn: Ferdinand Schöningh, 2006.

Schulmann, Robert, ed. *Seelenverwandte. Der Briefwechsel zwischen Albert Einstein und Heinrich Zangger (1910–1947)*. Zürich: NZZ Libro, 2012.

Schwabe, Klaus. *Wissenschaft und Kriegsmoral. Die deutschen Hochschullehrer und die politischen Grundfragen des Ersten Weltkriegs*. Göttingen: Muster-Schmidt, 1969.

Seelig, Carl. *Albert Einstein und die Schweiz*. Zürich: Europa, 1952.

———. *Albert Einstein. Leben und Werk eines Genies unserer Zeit*. Zürich: Europa, 1960.

———. *Albert Einstein. Mein Weltbild*. Frankfurt am Main: Ullstein, 1991.

Sexl, Roman and Herbert Kurt Schmidt. *Raum – Zeit – Relativität*. Braunschweig: Springer, 1991.

Sloterdijk, Peter. *Zur Welt kommen – Zur Sprache kommen*. Frankfurt am Main: Suhrkamp, 1988.

———. *Scheintod im Denken. Von Philosophie und Wissenschaft als Übung*. Berlin: Suhrkamp, 2010.

Sommerfeld, Arnold. "Zum siebzigsten Geburtstag Albert Einsteins." *Deutsche Beiträge* 2, München, 1949.

Sösemann, Bernd, ed. *Theodor Wolff. Tagebücher 1914–1919*. Erster Teil, Boppard: Oldenbourg, 1984.

Sösemann, Bernd, and Jürgen Frölich, eds. *Theodor Wolff*. Berlin: Hentrich & Hentrich, 2003.

Stachel, John, et al., eds. *The Collected Papers of Albert Einstein*, Vols. 1–13, Princeton: Princeton University Press, 1987–2015.

Staude, Jürgen, and A. Hofmann. "Sonnenforschung in Potsdam." In *300 Jahre Astronomie in Berlin und Potsdam*, edited by Wolfgang Dick and Klaus Fritze. Frankfurt am Main: Harri Deusch, 2000.

Stern, Fritz. "Freunde im Widerspruch. Haber und Einstein." In *Forschung im Spannungsfeld von Politik und Gesellschaft. Geschichte und Strukturder Kaiser-Wilhelm- / Max-Planck-Gesellschaft*, edited by Rudolf Vierhaus and Bernhard vom Brocke, 222–54. Stuttgart: Deutsche, 1990.

Stoltzenberg, Dietrich. *Fritz Haber*. Weinheim: Wiley-VCH, 1994.

Supf, Peter. *Das Buch der deutschen Fluggeschichte*. Vol. 2, Stuttgart: Drei Brunnen, 1958.

Szöllösi-Janze, Margit. *Fritz Haber*. München: C. H. Beck, 1998.

Tollmien, Cordula. "Der 'Krieg der Geister' in der Provinz – das Beispiel der Universität Göttingen 1914–1919." In *Göttinger Jahrbuch*, vol. 41, 137–209. Göttingen: 1993.

Trischler, Helmuth. *Luft- und Raumfahrtforschung in Deutschland 1900–1970*. Frankfurt am Main: Campus, 1992.

Troeltsch, Ernst. *Spektator-Briefe. Aufsätze über die deutsche Revolution und die Weltpolitik 1918/22*. Tübingen, 1924.

Ungern, Jürgen, and Wolfgang von Sternberg. *Der Aufruf. An dic Kulturwelt*. Stuttgart: Peter Lang, 1996.

Verhey, Jeffrey. *Der "Geist von 1914" und die Erfindung der Volksgemeinschaft*. Hamburg: Hamburger Edition, 2000.

Vierhaus, Rudolf, ed. *Das Tagebuch der Baronin Spitzemberg*. Göttingen: Vandenhoeck, 1963.

Vierhaus, Rudolf, and Bernhard vom Brocke, eds. *Forschung im Spannungsfeld von Politik und Gesellschaft. Geschichte und Struktur der Kaiser-Wilhelm- / Max-Planck-Gesellschaft*, Stuttgart: Deutsch Verlags–Anstalt, 1990.

Vom Brocke, Bernhard. "Wissenschaft und Militarismus." In *Wilamovitz nach 50 Jahren*, edited by William M. Calder III, Hellmut Flashar, and Theodor Lindken. Darmstadt: Wissenschaftliche Buchgesellschaft, 1985.

Von Bloch, Johann. *Der zukünftige Krieg in seiner technischen, volkswirtschaftlichen und politischen Bedeutung*. Berlin: Puttkammer & Mühlbrecht, 1899.

Von Laue, Max. *Gesammelte Schriften und Vorträge*. Braunschweig: Vieweg, 1961.

Von Moltke, Helmuth. *Gesammelte Schriften und Denkwürdigkeiten. Vol. 7 Reden des General- Feldmarschalls Grafen Helmuth von Moltke*. Berlin: E. S. Mittler un Sohn, 1892.

Von Suttner, Bertha. *Die Barbarisierung der Luft*. Internationale Verständigung, Heft 6. Berlin: Friedens-Warte, 1912.
Walter, Otto. *Bider, der Flieger*. Olten: Walter, 1938.
Wazeck, Milena. *Einsteins Gegner*. Frankfurt am Main: Campus, 2009. [*Einstein's Opponents: The Public Controversy about the Theory of Relativity in the 1920s*. Translated by Geoffrey S. Koby. Cambridge: Cambridge University Press, 2014.]
Wehler, Hans Ulrich. *Deutsche Gesellschaftsgeschichte 1914–1949*. München: C. H. Beck, 2003.
Wenzel, Gisela. "Schöneberg voran!" In *August 1914. Ein Volk zieht in den Krieg*. Berliner Geschichtswerkstatt. Berlin: Nishen, 1989.
Wilde, Harry. *Walther Rathenau*. Hamburg: Rowohlt, 1971.
Willstätter, Richard. *Aus meinem Leben*. Weinheim: Chemie, 1940.
Winkler, Heinrich August. *Weimar 1918–1933*. München: C. H. Beck, 1993.
Wuensch, Daniela. *Zwei wirkliche Kerle*. Göttingen: Termessos, 2005.
Wussing, Hans. *6000 Jahre Mathematik*. Heidelberg: Springer, 2009.
Zott, Reginna. *Fritz Haber in seiner Korrespondenz mit Wilhelm Ostwald sowie in Briefen an Svante Arrhenius*. Berlin: Ellen R. Swinne, 1997.
——. *Wilhelm Ostwald und Walther Nernst in ihren Briefen*. Berlin: Verlag für Wissenschafts -und Regionalgeschichte, 1996.
Zuelzer, Wolf. *Der Fall Nicolai*. Frankfurt am Main: Societäts, 1981.

Photography Credits

Image 1: Free Domain
Image 2: Free Domain
Image 3: © Max-Planck-Society Archive, Berlin-Dahlem.
Image 5: © akg/Science Photo Library
Image 6: © akg/ Science Photo Library
Image 7: © akg-images
Image 8: Free Domain
Image 9: © akg-images
Image 10: © IAM/akg-images/World History Archive
Image 11: Free Domain
Image 12: © akg-images

Index of Persons

L

M

N

O

P

R

S

T

V

W

Z